Life of Fred®

Logic

Life of Fred®
Logic

Stanley F. Schmidt, Ph.D.

Polka Dot Publishing

ISBN: 978-1-937032-55-5

Printed and bound in the United States of America

Polka Dot Publishing Reno, Nevada

To order copies of books in the Life of Fred series,

visit our website PolkaDotPublishing.com

Questions or comments? Email the author at lifeoffred@yahoo.com

Second printing

Life of Fred: Logic was illustrated by the author with additional clip art furnished under license from Nova Development Corporation, which holds the copyright to that art.

for Goodness' sake

or as J.S. Bach—who was
never noted for his plain
English—often expressed it:

Ad Majorem Dei Gloriam

(to the greater glory of God)

If you happen to spot an error that the author, the publisher, and the printer missed, please let us know with an email to: lifeoffred@yahoo.com

As a reward, we'll email back to you a list of all the corrections that readers have reported.

A Note to Readers

The first six chapters of this book make a dandy high school course in logic.

Adding chapters 7–16 makes a full college-level course.

This book is complete. It has many topics that other logic books omit.

★ Every logic book talks about the five connectives—&(and), ¬(not), ∨(or), ⇒(implies), and ⇔(if and only if)—but few reduce them all down to one connective that can do the job of all five.

★ Few present 17 fallacies of logic—ultimately convincing six-year-old Fred that he has a wife.

★ Many present Gödel's Incompleteness Theorems, but few prove the Diagonal lemma that is at the heart of those theorems.

When I'm writing, it is in this typeface. (Times New Roman)
When you, my reader, talk, ***it is in this typeface.* (*Allegro*)**
When Fred is thinking, he prefers AustinsHand.

Puzzles are scattered throughout the book. I hate the thought of "exercises" or "problems." If you want exercise, head to the gym. If you want problems, tell the government you are not going to pay taxes.

Puzzles are meant to feel more like . . . puzzles. Some are super easy, and some might stump your logic teacher.

PREREQUISITES FOR LOGIC

You won't need any algebra, geometry, trig, or calculus. That's the good news.

On the other hand, having hair under your arms (shaved or not) for a couple of years is a fair indication that your brain's reasoning power is developed enough to work with logic.

You won't need a calculator or a protractor or a computer. There are no separate teacher's manuals, answer books, or DVDs. It's all right here.

This book has some giggles . . . like when the Duck walks into Fred's office on the first page.

This book is cheap.* No other complete logic book can compete. I, your author, retired from high school and college teaching in 1980 and have no need for gobs of royalty money at this point to pay my light bills.

One last thing . . . hundreds of thousands of readers have asked for a picture of the author. On the back covers of many books, publishers like to stick photos of the author.

Usually, for male authors, the photo shows him standing in the wind with a leather coat—a really rugged guy. For female authors, she will be pictured as every man's sweetheart.

The publisher had decided that no *Life of Fred* book will have Stan's picture on the cover. He explained, "It might hurt sales."

However, he can't stop me from including a photo inside the book. Here is my photo taken several years ago. This really is me!

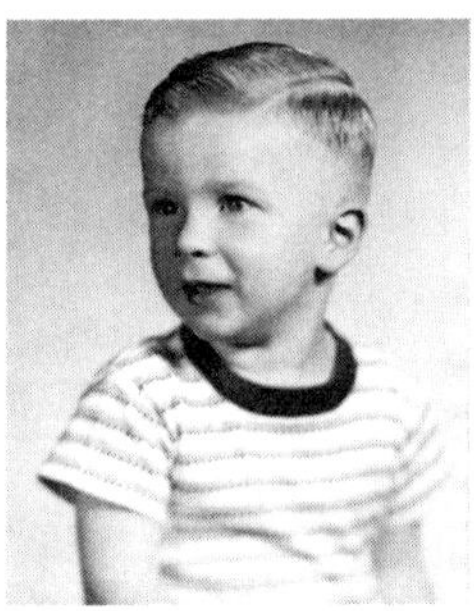

Dr. Schmidt

With my best wishes,

Stan

---

*The publisher wanted me to write *inexpensive*, but I like *cheap* better.

Contents

Chapter One
Sentences

A scratch at the door. It was almost like someone was rubbing feathers against the door. Fred looked up from his logic lecture notes that he had been working on. He hopped off his chair and headed to the door.

"I wonder who that could be," he said to his doll Kingie. "We usually don't have students visiting my office on Saturday afternoon."

Kingie shrugged his shoulders and continued working on his oil painting.

More scratching. Fred opened the door. His heart sank to the floor. It was the Duck.

The last time Fred had seen the Duck was a year ago when Fred was 5.* It had been a very bad experience for Fred.

Fred tried to never judge someone by their appearance. There could be a million reasons why Duck was wearing a tie, a sports coat, and sunglasses. The fact that Duck was 4½ feet tall and Fred was only 3 feet tall wasn't that important.

What did bother Fred were the sentences that Duck uttered:
"**Good morning.**" (It was 2 p.m.)
"**This is a lovely barn with lots of cows.**" (Kingie didn't like being called a cow, and this was Fred's math office. It wasn't a barn.)
"**You've grown a lot since the last time I saw you.**" (Fred has been 36 inches tall and 37 pounds for a long time. It was the Duck that had grown six inches in the last year.)

Every time Duck spoke, he lied. Every sentence was false.

---

*In the very first book of the Elementary Series: *Life of Fred: Apples.*

When Fred was 5, this really bothered him. In fact, back in *Apples* Fred ran away to escape from Duck. He couldn't stand hearing lie after lie.

Now at the age of 6, Fred's feelings toward Duck had changed. "Please come in and sit down."

Duck came in and said, "I'd rather stand," and then he sat down.

◇ ◇ ◇

Stop! Wait a minute. I, your reader of this book, have a question. I am grateful that you, Mr. Author, allow me to interrupt. Most other authors jabber and permit no one to ask for clarification.

What did you, my reader, have in mind?

Why in blazes did Fred invite Duck in? I would have turned him around and kicked him in the tail feathers. Duck just utters pure lies. If I wanted that, I would just turn on the television news.

There are three reasons why Fred welcomed Duck instead of kicking him out. ① Duck is 50% taller than Fred.* If you were six feet tall, that would be like messing with someone who is nine feet tall. That's NAGI (Not A Good Idea). ② From the third sentence at the start of this chapter, we note that Fred is working on his logic lecture notes. In his logic class on Monday—today is Saturday—Fred would love to bring Duck with him. Duck is a good example of *sentences* in logic. ③ Duck is a Fountain of Truth—I'll tell you about that later. Right now, I want to concentrate on ② and what *sentences* in logic are.

Go ahead. Who's stopping you?

Um. Didn't you interrupt me?

Gulp. Sorry. Go on with your story.

◇ ◇ ◇

Fred explained, "Sentences in logic are different than sentences in English. Sentences in logic must be either true or false."

*Duck is 4½ feet tall. That's 4.5 feet. Fred is 3 feet tall. Back from old *Decimals and Percents* days, if you wanted to compute "50% more than," you start with 100% and tack on another 50%. Then 50% taller than 3 feet means 150% of 3 feet. In decimals this means 1.5×3, which is 4.5.

Kingie set down his paint brush and complained, "What makes Duck so special? It's unfair that you are thinking of taking him to your logic class and not me. Humpf! Take me!"*

Fred smiled. "It's simple. Every sentence Duck says is a sentence in logic—it is either true or false. Every sentence you have just said is not a logic sentence.

✓ "Your first sentence—What makes Duck so special?—is a question.

✓ "Your second sentence—It's unfair . . .—is an opinion.

✓ "Your third sentence—Humpf!—is an interjection.

✓ "Your fourth sentence—Take me!—is a command, which in some English classes is called an imperative.

"None of your English sentences is a logic sentence."**

Duck wanted to show off and said:

Kingie is a ten-pound elephant.
All rabbits are white.
There is a state in the U.S.A. that begins with the letter B.

These are three sentences in logic. Fred wrote them down in his logic notes and labeled them K, R, and B.

Every sentence in logic is abbreviated with a capital letter.

---

*Small quick explanation for those readers who are new to the *Life of Fred* series: Kingie is Fred's doll. They have known each other for almost all of Fred's life. When Fred was four days old, the man at King of French Fries gave this doll to Fred. That's why Fred named his doll Kingie.

Fred didn't have to tell Kingie that he was thinking of taking Duck to his logic class on Monday, because, as everyone knows, dolls can read their owners' minds.

**In English when you have several paragraphs quoting the same person, only the last paragraph has the close quote (") symbol.

And while we're doing English, please note that we don't write, "None of your English sentences *are*. . . ." The subject of the sentence is *none*, and so the verb is singular—*is,* not *are*.

Kingie was not in a good mood. In the past, when Fred had a cat, dogs, or a llama as pets, they were all bad news for Kingie. Now Fred was bringing a lying duck into his office. Kingie muttered, "Logic is nuts. (opinion) What if you have 27 logic sentences? (question) You are going to run out of letters! (a real logic sentence)" Kingie wanted to have the "last word" before he headed back to painting.

Duck repeated Kingie, "**You are going to run out of letters!**"

Kingie was fuming. He understood what it means when a duck that always lies *agreed* with him? Kingie put a new canvas on his easel and painted the ugliest duck he could.

Fred didn't know what to do in this emotionally charged situation. He knew that on Monday when he introduced the sentence letters ABCDEFGHIJKLMNOPQRSTUVWXYZ, that someone might ask about the case in which you have 27 logic sentences.

He would first say that it was quite rare in logic to have more than three sentences to play with. In most cases, logicians liked to use A, B, and C, or P, Q, and R.

Years ago when Fred was first teaching logic, he used to say that if you had more than 26 logic sentences, you could use letters from other alphabets: ΔΦΓϑΛΠΘΣΩΞΨ, or БДЖИЛЦЯЮЂҖ, or שקץעסמלחההדבא. But some student would then ask, "Well, what if you had a thousand logic sentences? You would run out of different alphabets."

Fred was older (age 6) and wiser now. When asked that question on Monday, he would say that logicians had solved that puzzle. If they needed a lot of logic sentences, instead of using ABCDEFGHIJKLMNOPQRSTUVWXYZ, they use P_1, P_2, P_3, P_4. . . . Logicians have an infinite number of sentence letters.

It's tough to win a math argument with Fred.

Fred liked the fact that Duck could only utter logic sentences, but that's not what intrigued Fred. He realized that Duck was a Fountain of Truth.

That's nuts! I, your reader, know that's the sun and not a cheeseburger. How in the world can Fred think that this lying Duck is a Fountain of Truth ***as he calls it?***

Puzzle #1: If you want to find out from Duck the truth about something, you first have to learn whether Duck is knowledgeable about that topic. For example, if you want to know whether the Goldbach conjecture is true, you can be fairly certain that Duck doesn't know the answer.*

Note to readers:
Answers to all of the puzzles are given in the back of this book.
Please do not just read the question and turn to the answer. That would be like going to a gym and just watching people work out.

KITTENS University is where Fred teaches. It is in Kansas. Suppose that Fred wanted to find out *whether Duck knew* what the capital of Kansas is. What question could Fred ask Duck?

---

*That's because no one (today) knows the answer.

Back in 1742 Christian Goldbach had two conjectures. The first was that every even integer greater than 2 can be written as the sum of two primes. $4 = 2 + 2$ $6 = 3 + 3$ $8 = 5 + 3$ $10 = 7 + 3$ and so on.

His second conjecture was that every integer greater than 5 can be written as the sum of three primes. $6 = 2 + 2 + 2$ $7 = 2 + 2 + 3$ $8 = 2 + 3 + 3$ and so on.

For over 250 years no one has been able to figure out whether either of these conjectures is true.

Once you have determined that Duck knows what the capital of Kansas is, the next step is to force this lying duck to become a Fountain of Truth.

You can't just ask Duck, "What is the capital of Kansas?" He will lie and say that Hot Dog is the capital.

If you don't know what the capital of Kansas is, you can't just keep making guesses: Is Sacramento the capital of Kansas? Is Bismark the capital of Kansas? Is Berlin the capital of Kansas? That would take forever.

Puzzle #88: [harder] What question could you ask Duck in order to force him to tell you that Topeka is the capital of Kansas? (The puzzles are not numbered consecutively so that you won't accidently see the answer to this question when you read the answer to Puzzle #1.)

Note to reader:

In many parts of math—such as algebra—it is important that you know the current material before you head on to the next chapter. For example, if you don't know how to factor trinomials such as $x^2 - 5x + 6$, *then solving* $(x^2 - 5x + 6)(x^2 + 6x + 8) = 0$ *can be a real pain.*

In contrast, in this course in logic, these puzzles can be a source of pleasure for a week. It might be next Monday before you figure out how to squeeze "Topeka" out of Duck.

The Two Halves of Logic

Pure Logic

Logic has a language. We will call it L.

So far, we know that L contains sentence letters such as A, B, C, or P_1, P_2, P_3, P_4.

Applied Logic

Here is where we *use* that logic language L.

We create a **model** for L in which every sentence letter has a meaning. This is called **semantics**.

B might stand for "Betty is a Ph.D. student at KITTENS.

C might stand for "Fred is a canary."

One model for L can be set theory, which we will look at in Chapter 8.

Another model for L can be arithmetic, which we will look at in Chapter 9.

In fact, virtually all parts of math (including geometry) can be thought of as models of L.

Models of L are sometimes called **structures** for L.

Memory Aid Tie these four words together:

model,

meaning,

semantics, and

structure.

They are all part of applied logic.

Duck is a semantical kind of bird. All we get out of him are models for L. For example:

A is Apples sing in the moonlight.
B is Bowling balls are my favorite candy.
C is Cinderella likes eating toothpaste on toast.

In pure logic, in language L, we just have sentence letters like P, Q, and R. It would be silly to ask, "Is P true?"

It is only when we assign a structure to L, that we can discuss whether P is true or false.

When Fred teaches logic, he has two favorite models* he likes to talk about: set theory and arithmetic.

Hold it! Stop the show! I, your reader, am starting to panic. I don't care if Fred messes with arithmetic. I can handle 2 + 3 = 5, but it's been a hundred years since I've done any set theory. The only thing I can remember is that a set is any collection of objects. That's it. Before you, Mr. Author, go any further, tell me everything about set theory.

---

*models = structures

Everything? Geep! There are *books* written about set theory. There are mathematicians who spend their whole lives just playing with sets. I can't . . .

I don't mean everything about set theory. How about just a bit of a refresher. Just the super basics. I bought this book, and I want you to follow my wishes.

I thought authors were supposed to figure out what goes in a book.

This is a brave new world. Without me, you are nothing.

Wow. That sounds like solipsism. Some of my readers already know about sets. I'll put the basics about sets in a box on this page. Then those other readers can skip over the box if they want to.

It's a deal.

Handy Short Course in Set Theory

A set is any collection.

{⏲, ❀, ✈} is a set that contains three members.

"{" and "}" are called braces. You stick braces around a set.

∊ means "is a member of." ❀ ∊ {⏲, ❀, ✈}

∉ means "is not a member of." ☎ ∉ {⏲, ❀, ✈}

ℕ is the set of natural numbers {1, 2, 3, 4, 5, . . .}.

¾ ∉ ℕ

Two sets are equal if they have exactly the same members.

If A and B are equal sets, then we write A = B. This is a different "=" than the one used in arithmetic.

Another word for *members* is *elements.*

The world's smallest set is { }, which is called the empty set. It has no elements in it.

The cardinality of a set is the number of elements in the set. The cardinality of {⏲, ❀, ✈} is 3. The cardinality of { } is 0.

That amount of set theory should hold us for now. On Monday, if Fred wanted to create a set theory structure for L, all he would need to do is assign each sentence letter of L to some true-or-false sentence in set theory.

Show me!

Fred could assign A to ❀ ∈ {⌚, ❀, ✈}.
He could assign B to ❀ ∉ {⌚, ❀, ✈}.
He could assign C to {⌚, ❀, ✈} = {⌚, ❀, ✈}.

Ha! I caught you. There are an infinite number of sentence letters in L. ***There are*** C, D, E, . . . , ***and there is the infinite list*** P_1, P_2, P_3, P_4. ***He could spend all Monday and never finish making his model.***

You didn't let me finish. I was going to say that Fred would then assign all other sentence letters of L to ✈ ∈ {⌚, ❀, ✈}.

Often in making a model, we will only need to assign some of the sentence letters. The rest of the sentence letters can be assigned to anything in the model. It is a little like throwing the unneeded sentence letters in the trash can.

Puzzle #40: Create an *arithmetic* model for L.
Here's a start. Assign A to ¼ + ⅔ = 27.

I, your reader, hate to interrupt again**, but this pure logic thing seems so stupid. All you have are sentence letters—like A, B, or C—and there's nothing to do with them. They can't even be true or false in pure logic.

Fred enjoys juggling them.

Before you interrupt again, I would like to point out that this is only Chapter 1 in which I've introduced sentence letters. Once we get to Chapter 2 and introduce connectives, things will get much more interesting in pure logic.

How soon do we get to Chapter 2? I'm ready!

Just turn the page.

---

*This is obviously a lie. Duck is having a bad influence on you. You seem to enjoy the freedom you have to interrupt me.

Chapter Two
Connectives

When Duck told Fred that he wasn't the slightest bit hungry, then Fred knew that it was very important to find some food for him. Duck was starving. Fred opened up his desk drawers where he had stored food "for later" and offered it to Duck.

Duck said, "**I don't want any of your food**," and then began to devour it saying, "**This tuna sandwich is yucky**" and "**This bunch of grapes tastes terrible.**" Fred knew that Duck was very happy.

> ### *Intermission*
>
> While Duck is gorging himself, it's time to admire that lying bird. Here is why. If you add "It is not the case that . . . " in front of whatever Duck says, then you have pure truth.
>
> If Duck says, "Everybody likes anchovies in their milkshakes," then "It is not the case that everybody likes anchovies in their milkshakes" is true.
>
> The truly evil liar is the one that mixes truth with falsehood. Most of what they say is true, but inside their talk is hidden the poisonous lie.
>
> Example #1: Someone is trying to talk you into using heroin. They say:
>
> 1. The first sample is free. (true)
> 2. It will make you feel good. (true, at least temporarily)
> 3. The stick in the arm doesn't hurt much. (true)
> 4. You can always quit. (not if it kills you)
>
> Example #2: Most of what Adolf Hitler said was true. He would tell the crowds that they were a special people, that their achievements in science, art, and literature were remarkable, that blood and soil (fighting and agriculture) are a deep part of the German culture, that it would be nice to have some more "living space." (← all true)
>
> Then he would draw the non-logical conclusion that we "must" kill off other people and take their lands and possessions.

"It is not the case that . . ." is our first connective. If the sentence letter A represents (in some model) "Everybody likes anchovies in their milkshakes," then we know that A is false.

The symbol we use for "it is not the case that . . . " is " ¬".

¬A is true.

If the sentence letter K (in some structure) is "Topeka is the capital of Kansas," then K is true. And ¬K is false.

Here is a handy little chart for the ¬ connective:

P	¬P
T	**F**
F	**T**

Pure logic is easy.

Applied logic with its structures is both harder . . . and more useful.

In pure logic you can say B or you can say ¬B. In applied logic you enter the messy world of models where you often have to use **English**. Everyone knows that **English** is harder than math. In logic you say B. In some model it might be:

That bird is blue. —or—
The color of that bird is blue. —or—
Blue is the color of that bird. —or—
People who are not color blind think that bird is blue.

Puzzle #56: How many different ways in English could you express "If Kingie paints a picture, then someone will buy it." As usual, please don't just read this puzzle and turn to the answer. Learning how to think is much more valuable than seeing the nine (!) ways I thought of to answer this puzzle.

The ability to do sustained thinking—that's a thing this world values highly.

Babies (and people who just read the question and then look at the answer) are good for four seconds of thought. Advanced babies might spend a whole minute playing with a rattle.

What do you want in life AND how long and hard are you willing to work for it?

Our language L in pure logic originally consisted only of sentence letters such as A, B, C, . . . P, Q, R, and P_1, P_2, P_3, P_4. . . .

We are now going to expand L with the connective ¬.

Definition:
1. Every sentence letter is a **sentence** in L.
2. If P is a sentence in L, then so is ¬P.

Hey! I, your reader, think this type of definition sounds vaguely familiar. It smells like something we did back in algebra. It feels and tastes like what I've seen before, but I don't see it quite yet.

Wow. You really have all five senses working. This kind of definition is called a **recursive definition**. This may be the first recursive definition you have ever encountered.* You'll see several more in this book.

A recursive definition has two parts. If you are recursively defining, say, what a ЖэѿЂ is, you first say that certain things you already know are ЖэѿЂ. In the second part, you build on the ЖэѿЂ that you already know in order to get new ЖэѿЂ.

We can recursively define the natural numbers ℕ = {1, 2, 3, . . .}:
1. 1 is a natural number.
2. If x is a natural number, then x + 1 is also a natural number.

I know what definitions by recursion reminds me of! Back in advanced algebra we did proofs by induction. If we wanted to prove a whole list of statements S_1, S_2, S_3, . . . were true, we first proved that S_1 was true. Then in the induction step we showed that if S_n were true, then S_{n+1} would also have to be true.

Beautiful. Recursive definitions and proof by induction are like cousins. One is used to define things, and one is used to prove things.

Puzzle #7: Using the recursive definition of a sentence in L (see the top of this page), show that ¬¬¬B is a sentence in L.

---

*Assuming you have not read *Life of Fred: Five Days of Upper Division Math.*

Before that recursive definition, language L had an infinite number of elements: A, B, C, . . . P, Q, R, and P_1, P_2, P_3, P_4. . . .

Now each of those elements, such as C, now have an infinite number of "babies": ¬C, ¬¬C, ¬¬¬C, ¬¬¬¬C. . . .

Lots and lots of sentences in L—and we've just begun.

In any particular model of L, C would have a definite truth value. It would either have to be true or it would have to be false. Say, for example, that C was false.

Then ¬C would be true.

Then ¬¬C would be false.

Then ¬¬¬C would be true. etc.

In one model, C might be . . .

C is false.

¬C ("Cheesecake doesn't taste better with gasoline sauce") is true.

The Point: The recursive definition of a sentence in L "works." I.e.,* if all the sentence letters of L (A, B, C, . . . P, Q, R, and P_1, P_2, P_3, P_4 . . .) are either true or false in any particular model, then so must all the sentences of L (¬A, ¬¬¬Q, ¬¬P_{47}, . . .) be also be true or false in that model.

But you probably knew that already. That little truth table from two pages ago spelled that out.

	P	¬P
Here's one possible structure ➸	**T**	**F**
Here's another possible structure ➸	**F**	**T**

In any particular model, P must be either true or false.

*I.e. = "that is to say"

¬ attaches to a single sentence in our language L.

All the other connectives attach to a pair of sentences. Here is the list . . .

&	both . . . and . . .	"and"
∨	either . . . or . . . or both	"or"
⇒	if . . . then . . .	"implies"
⇔	. . . has the same truth as . . .	"iff"*

Here is a **complete discussion** of each of these connectives.

And You can find & as a "capital 7" on most keyboards.

P & Q is true only when both P and Q are true.

In semantics** there are four possible models for P and Q.

case 1: P is true; Q is true.
case 2: P is true; Q is false.
case 3: P is false; Q is true.
case 4: P is false; Q is false.

Each model is a line in the truth table:

P	Q	P & Q
T	**T**	**T**
T	**F**	**F**
F	**T**	**F**
F	**F**	**F**

Truth tables are a lot easier than writing out in English:

case 1: P is true; Q is true. In this case P & Q is true.
case 2: P is true; Q is false. In this case P & Q is false.
case 3: P is false; Q is true. In this case P & Q is false.
case 4: P is false; Q is false. In this case P & Q is false.

---

*iff = if and only if. The word *iff* is in most good dictionaries. P ⇔ Q means both P ⇒ Q and Q ⇒ P.

Recall the **Memory Aid ties together *models*, *meaning*, *semantics*, and *structure* as parts of applied logic.

The only real problem with the connectives & ("and"), $\vee$ ("or"), and $\Rightarrow$ ("implies") is . . . not with pure logic

. . . not with applied logic

. . . but with **English**.

And in English has more than one meaning. A & B in logic has the same truth as B & A. But in English, these two sentences are not the same: They had a baby and got married.

They got married and had a baby.

And in English can have the meaning *and then.*

Sometimes in English *and* is replaced by nothing. If in some model S means "Fred is short," and T means "Fred is a teacher," then S & T in logic might be expressed in English as "Fred is a short teacher."

Or P $\vee$ Q in logic will be true if either P is true or Q is true or both of them are true. The only time P $\vee$ Q will be false is when P is false and Q is false.

Each model of L will be a line in the truth table:

P	Q	P $\vee$ Q
T	**T**	**T**
T	**F**	**T**
F	**T**	**T**
F	**F**	**F**

In a structure of L in which P is true and Q is false, then P $\vee$ Q will be true.

$\vee$ in logic corresponds to the **inclusive or** in English. (= one or the other or both) This is probably the rarer meaning of *or* in English. If you say that every kid was prepared for a long wait in the bus station because each of them brought either a smart phone or computer tablet, you are using the inclusive *or*, you wouldn't be lying if some kid brought both.

However, in many everyday uses of *or* in English the **exclusive or** is used. The policeman yells, "Stop or I'll shoot!" You are hoping that he is using the exclusive *or*. You expect that only one of those two alternatives is true. If you stop and got shot, you would shout "unfair!"

When my daughter was about 4, she pretended she didn't understand the exclusive *or* in English. I asked her, "Would you like ice cream or cherry pie?" She smiled and said, "Yes." She was using the logic ∨—the inclusive *or*.

Where did the symbol ∨ come from? You may have already guessed, but let me mention its origin for the 99.9993% of the readers who haven't guessed. ∨ comes from the Latin word *vel*, which is the inclusive form of *or* in Latin. A ∨ B when at least one of them is true.

I, your reader, have a different quesiton. I like to be prepared for the future. I know that ¬ is used in front of a single sentence in* L*. You have told me that we have four connectives—*&, ∨, ⇒, *and* ⇔ *that will attach to a pair of sentences.

My question is will we get connectives that attach to three sentences?

Why don't you just ask Duck?

I am relieved.

The logical connectives are a little like the connectives in arithmetic. There's the minus sign that changes 6 into –6. There are the addition and the multiplication signs. They only attach to pairs of numbers.

You never add three numbers *simultaneously.* If you are working on 5 + 3 + 7, you add two of the numbers together and then add the third number. 5 + 3 + 7 means (5 + 3) + 7, which means 8 + 7.

That means that all I have yet to learn is ⇒ and ⇔! That's swell.

Swell? You have been watching too many movies from the 1940s. Who nowadays says *swell*? And *keen* is also out of fashion.

Okay. I think it is great, wonderful, marvelous, first-rate that the only connectives left to learn are ⇒ and ⇔.

Puzzle #70: Let's do a little counting. How many possible logical connectives could be placed between two sentences in logic? To review for a moment: The **fundamental principle of counting** states that if there are *m* ways to do one thing and *n* ways to do another, then there are *mn* ways to do both. If there are six possible salads and nine possible pizzas, then there are 54 possible happy dinners.

To answer the question of how many possible connectives can be placed between P and Q, we have to answer four questions. For fun, let's say that ✱ is a possible connective. Our first question is in the case that P and Q are both true. Then will P ✱ Q be true or false? The second question is in the case that P is true and Q is false. Then will P ✱ Q be true or false?

Implies The only time P ⇒ Q will be false is when P is true and Q is false. In logic we don't want to start out with truth (P) and end with something false (Q).

from this

In logic we don't want to go

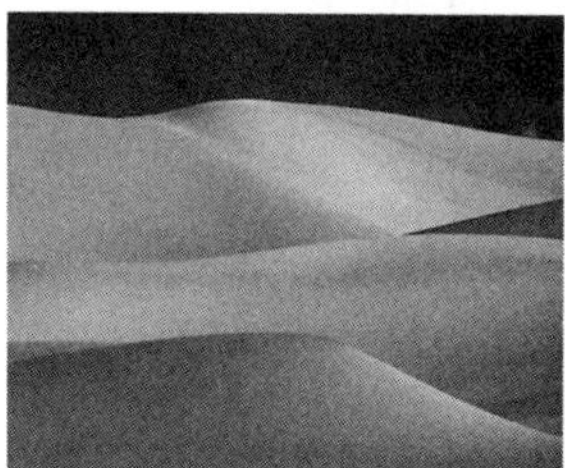

to this.

No one (except Duck) would argue with the first two lines in the truth table for implies.

P	Q	P ⇒ Q
T	**T**	**T**
T	**F**	**F**

The real question how to handle P ⇒ Q when P is false.

Logicians claim
that when P is false
P ⇒ Q is always true.

Here's why **F** ⇒ **F** is true:

Your mother tells you, "If you are nice to Aunt Nancy, I will double your allowance." She is saying N ⇒ D.

Suppose you are really snotty to Nancy. Suppose your mother doesn't double your allowance. We are in the situation where N is **F** and where D is **F**.

Was your mother lying to you when she said N ⇒ D? Of course not.

Here's why **F** ⇒ **T** is true:

Your father tells you, "If you snort more than four pounds of cocaine a week, you will wind up in the hospital." C ⇒ H is a true statement.*

Suppose you don't use cocaine at all. Then C is **F**. Suppose you wind up in the hospital (because you were speeding on your motorcycle—the one with bald tires—on a winding road at midnight in a snow storm and you were drunk). In this situation we have **F** ⇒ **T**. Was your father lying to you when he told you that using more than four pounds of cocaine a week would put you in the hospital? No, he wasn't.

Here is the complete truth table for implies:

P	Q	P ⇒ Q
T	**T**	**T**
T	**F**	**F**
F	**T**	**T**
F	**F**	**T**

Puzzle #32: Logicians say which of these (if any) are true?

A) If the moon is made of cheese, then mice made the craters.

B) If 2 + 2 = 4, then 785 = 785.

C) If Elvis was the first president of the U.S. then King Tut was left handed.

D) If King Tut was left handed, then Washington was the first president of the U.S.

---

*It's also true that if you snort that much cocaine, your nose will fall off.

Just for fun, here's a second reason why **F** $\Rightarrow$ **F** is true:

In logic we really like P $\Rightarrow$ P for any possible situation (any possible structure for L). P $\Rightarrow$ P just feels good.

If that is the case, then we will have to say that **F** $\Rightarrow$ **F** is true.

In short, A $\Rightarrow$ B will be true if and only if either A is false or B is true (or both).

Puzzle #62: Using only ¬, &, and ∨, express A $\Rightarrow$ B.

Iff P $\Leftrightarrow$ Q means that P and Q are both true, or it means that P and Q are both false. They are said to be **logically equivalent**.

Here is the truth table for iff:

P	Q	P $\Leftrightarrow$ Q
T	**T**	**T**
T	**F**	**F**
F	**T**	**F**
F	**F**	**T**

Using the results of Puzzle #62: (A $\Rightarrow$ B) $\Leftrightarrow$ (¬A ∨ B).

When we write definitions in mathematics, we use iff. For example, in geometry we say that quadrilateral* ABCD is a parallelogram iff its opposite sides are parallel.

That means two things: (1) if it's a parallelogram than its opposite sides are parallel, and (2) if the opposite sides are parallel, then it's a parallelogram.

Before you can object . . .

Wait a minute! How do you know what I, your reader, am thinking?

You might say that I can read your mind. The truth is that I've been teaching logic long enough that I know what you'll be objecting to.

But you are denying me my First Amendment rights!

*quadrilateral = a four-sided figure

Your rights under the First Amendment to the Constitution of the United States include your right to (1) peaceably assemble, (2) petition the government, (3) freely exercise your religion [as long as it doesn't include going around and killing others whom you consider heretics] and (4) enjoy freedom of speech and of the press.

So you gotta let me object!

Okay.

You made me forget what I was going to say!

May I? You were going to say that in many math books, the authors write their definitions using "if" instead of "iff." They will write `a triangle is isosceles if at least two of its sides are congruent.`

Exactly!

We need to forgive them. If you really need to, you can take a sharp pencil and change `is isosceles if at least` into `is isosceles if`$_{\wedge}^{f}$` at least.`

In English, we do a lot of confusing $\Rightarrow$ with $\Leftrightarrow$ and with $\Leftarrow$. You mother tells you, "If you are Housebroken,* then you can wear big people's Clothes (no diapers)." She's *saying* if H then C.

What you are *thinking* . . .

What she is *thinking* . . .

She says the happy thought (diaper freedom), but she's really thinking the negative thought (diaper jail). The truth is H $\Leftrightarrow$ C.

---

*potty trained

Puzzle #35: Suppose we have a model of our logic language L in which H is assigned to the sentence: "Next Tuesday a 100-megaton atomic bomb will be detonated over KITTENS University," and in which C is assigned to: "Some people at KITTENS will never reach their 60th birthday."

All other sentences of L are assigned to: "Some people think carrots are orange."

Your question: Is this a model for L?

Puzzle #93: (Continuing the previous puzzle) Which, if any, of these are true: H ⇒ C, C ⇒ H, or H ⇔ C?

So far in our study of sentence logic* we have the sentence letters A, B, C, . . . , P, Q, R, . . . , X, Y, Z, P_1, P_2, . . . and we have the connectives ¬, &, ∨, ⇒, and ⇔. We only have to add parentheses in order to have all the symbols needed for sentence logic.

Here's why we need parentheses:

Without parentheses A ∨ B & C has two different meanings. Suppose we have a structure for L in which:

A is assigned to "Able will be at KITTENS next year."
B is assigned to "Bob will be at KITTENS next year."
C is assigned to "Carrie will be at KITTENS next year."

A ∨ B & C *could mean* that either Abel or Bob will be at KITTENS, and Carrie will certainly be at KITTENS. (A ∨ B) & C

Or it *could mean* that either Able will be at KITTENS or Bob and Carrie will both be at KITTENS. A ∨ (B & C)

On page 26 we had a recursive definition of a sentence in L when we only had one connective:

Definition:
1. Every sentence letter is a **sentence** in L.
2. If P is a sentence in L, then so is ¬P.

We extend that definition for our new connectives.

---

*Fancy people call it **sentential logic**. Extra fancy people call it **propositional logic**.

Definition:
1. Every sentence letter is a **sentence** in L.
2. If P is a sentence in L, then so is $\neg$P.
3. If P and Q are sentences in L, then so is (P & Q).
4. If P and Q are sentences in L, then so is (P $\vee$ Q).
5. If P and Q are sentences in L, then so is (P $\Rightarrow$ Q).
6. If P and Q are sentences in L, then so is (P $\Leftrightarrow$ Q).

Puzzle #50: Using this recursive definition, establish that ((A $\vee$ B) $\Rightarrow$ (C & B)) is a sentence in L.

Puzzle #18: [harder] Argue from the definition of $\Rightarrow$, which has this truth table

P	Q	P $\Rightarrow$ Q
T	**T**	**T**
T	**F**	**F**
F	**T**	**T**
F	**F**	**T**

that A $\Rightarrow$ B and B $\Rightarrow$ A, must mean that A $\Leftrightarrow$ B.

Puzzle #78: Suppose we have a structure for L in which sentence letter P is assigned to: "Joe's favorite natural number is even." The natural numbers = {1, 2, 3, 4, . . .}.

And suppose that Q is assigned to: "If you square Joe's favorite natural number, the answer is even."
Which of these is/are true? P $\Rightarrow$ Q, Q $\Rightarrow$ P, and P $\Leftrightarrow$ Q?

English is one of the hardest parts of logic. When Fred was three months old, his mother wanted to tell him the important things that he should know for his adult life. Fred was expecting a Facts-of-Life lecture about the birds and the bees, but that wasn't forthcoming. In fact, she wasn't sure what the important things of life are. She thought for a couple of minutes and told him, "Drink 16 milkshakes in 10 minutes and you will regret it." (Her uncle had done that once.) This was the most significant piece of information that his mom had ever told him.*

If M is "You drink 16 milkshakes in 10 minutes," and R is "You will regret it," what she *meant* was not M & R, but M $\Rightarrow$ R. *And* in English can mean *implies.*

---

*See *Life of Fred: Calculus Expanded Edition* for all the other things she said to him.

Chapter Three
Making L Simpler

Fred wanted to make the sentence logic language L as simple as possible for his lecture on Monday. He remembered that Albert Einstein once said, "Everything should be made as simple as possible."* Fred was rewriting his lecture notes.

Logic language L was pretty simple to start with. L just consists of sentence letters like A, B, or P_3, some connectives (¬, &, ∨, ⇒, and ⇔) and a couple of parentheses and the truth tables for each of the connectives.

P	Q	¬P	P & Q	P ∨ Q	P ⇒ Q	P ⇔ Q
T	**T**	**F**	**T**	**T**	**T**	**T**
T	**F**	**F**	**F**	**T**	**F**	**F**
F	**T**	**T**	**F**	**T**	**T**	**F**
F	**F**	**T**	**F**	**F**	**T**	**T**

The first thing that Fred could get rid of in L was ⇔. He didn't need that connective. Every time you encountered A ⇔ B you could replace it with (A ⇒ B) & (B ⇒ A).

His list of connectives was down to four: ¬, &, ∨, and ⇒.

Next, he eliminated ⇒. Replace A ⇒ B with ¬A ∨ B.**

His list of connectives in L was down to three: ¬, &, and ∨. He called these three the three Musketeers. They could handle anything that the original five connectives could handle.

Then Fred realized that there was a big hole in his argument. It was true that ¬, &, and ∨ could handle any of the original five (¬, &, ∨, ⇒, and ⇔), but could they handle all 16 possible connectives? [Puzzle #70 on page 31]

---

*What Einstein really said was, "Everything should be made as simple as possible, but not simpler."

**We did that in Puzzle #62 on page 33.

Fred thought briefly (0.000037 seconds) and smiled. Of course he thought. Suppose P ✱ Q were given by the truth table

P	Q	P ✱ Q
T	T	F
T	F	T
F	T	F
F	F	F

Then I would want P ✱ Q to be true only when P is true and Q is false. So P ✱ Q could be represented by (P & ¬Q).

If P ❀ Q were given by the truth table

P	Q	P ❀ Q
T	T	T
T	F	F
F	T	T
F	F	T

then I would want P ❀ Q to be true in three different cases.

case 1: when P is true and Q is true.
case 2: when P is false and Q is true.
case 3: when P is false and Q is false.

case 1: (P & Q)
case 2: (¬P & Q)
case 3: (¬P & ¬Q)

Combining those three cases:
P ❀ Q would be (P & Q) ∨ (¬P & Q) ∨ (¬P & ¬Q).

So my three Musketeers (¬, &, and ∨) could represent any connective between TWO sentences.

Back on page 30, Mr. Author gave a phony argument why we would never need connectives between THREE sentences. Betty and Alexander will be in my logic class on Monday, and they won't accept the weak argument that just because we can't add three numbers together simultaneously, that really doesn't prove that some connective for three sentences doesn't exist—some connective that can't be represented by ¬, &, and ∨.

Here's what I can tell Betty and Alexander on Monday:

Suppose we have some three-sentence connector ⊕(P, Q, R) with a truth table

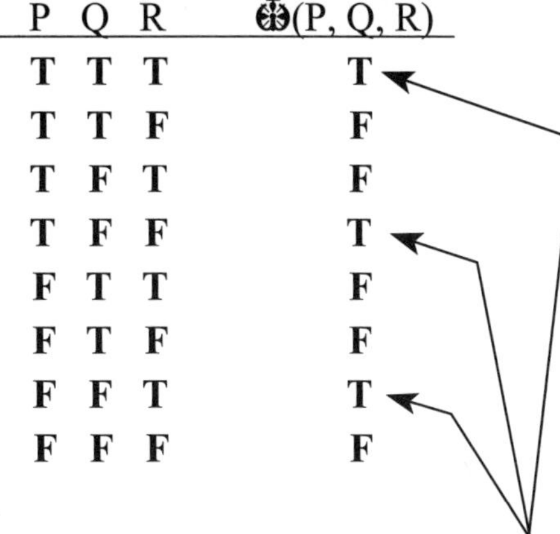

P	Q	R	⊕(P, Q, R)
T	**T**	**T**	**T**
T	**T**	**F**	**F**
T	**F**	**T**	**F**
T	**F**	**F**	**T**
F	**T**	**T**	**F**
F	**T**	**F**	**F**
F	**F**	**T**	**T**
F	**F**	**F**	**F**

I'll want ⊕(P, Q, R) to be true in three cases. So I can represent ⊕(P, Q, R) by (P & Q & R) ∨ (P & ¬Q & ¬R) ∨ (¬P & ¬Q & R).

And I can do the same with any four-sentence connector mutatis mutandis.*

I objec . . .

Before you can object and say that (P & Q & R) is not proper, please let me explain. By the recursive definition of sentences in L, which was given three pages ago, Fred should have written ((P & Q) & R).

Four pages ago I carefully noted that A ∨ B & C had two different meanings, depending on where you put the parentheses.

On the other hand, it doesn't matter where you put the parentheses when you have **multiple conjunctions** such as P & Q & R.
(A conjunction is a fancy name for &.) Logicians often leave out parentheses in this case. P & Q & R is easier to read than ((P & Q) & R).

Ditto for **multiple disjunctions**. (A ∨ B ∨ C) is easier to read than ((A ∨ B) ∨ C). Too many parentheses may cause temporary blindness, and we wouldn't want that.

---

*Fred sometimes starts to think in Latin. *Mutatis mutandis* is a common expression among mathematicians. It's also in most good English dictionaries. *Mutatis mutandis* means "after making the necessary obvious changes."

And while we are in the mood, we note that logicians when looking at some big logic sentence, such as

$$(((P \Rightarrow Q) \& (Q \Rightarrow R)) \Rightarrow (P \Rightarrow R)),$$

will often omit the *outermost pair* of parentheses

$$((P \Rightarrow Q) \& (Q \Rightarrow R)) \Rightarrow (P \Rightarrow R).$$

It makes things just a little easier to read.

When robots read A ⇒ B, they go crazy and exclaim, "`This is not a logical sentence by the recursive definition!`" When humans read A ⇒ B, they don't need the extra tinsel decoration that (A ⇒ B) has.

Puzzle #44: On the previous page you saw the eight lines needed to list all the possible **T**s and **F**s when dealing with three letters. Two pages ago you saw the four lines needed to list all the **T**s and **F**s for two letters.

You can guess what's coming. List the 16 lines needed to list all the possible **T**s and **F**s when dealing with four letters.

Your first line will be **T T T T** . Your last line will be **F F F F** .

Fred now knew that his three Musketeer connectives, ¬, &, and ∨, were all he needed to represent any model of L, regardless* of whether it had just two sentence letters or 20.

In fancy language, ¬, &, and ∨ are **expressively complete**.

Three—as in three connectives—seemed like a large number to Fred. *Do I need all three?* he thought.

What does P ∨ Q mean? It means that not both P and Q are false. That's the only situation in which P ∨ Q is not true.

So P ∨ Q can be written as ¬(¬P & ¬Q).

In fancy language, ¬ and & are expressively complete.

Fred just killed one of the three Musketeers.

I know.

You interrupted me. I was about to ask a question.

Sorry. Ask your question.

**Irregardless* ain't a word. If you use it, you sound dum.

That's a second interruption. Before you apologize for your two interruptions, let me ask a simple question. Is Fred going to kill one of the two remaining Musketeers—either ¬ or &—so that we are down to just either ¬ or &?

No. It's true that you can't express ¬P just in terms of &. Nor can you express P & Q just in terms of ¬.

Hurray! I can write in my notes that the minimum number of connectives for pure logic is 2.

No.

What do you mean? You just told me that you can't kill either ¬ or &. If you can't eliminate either of them, then the minimum is 2.

No.

Stop it! Stop just saying no.

I would be happy to explain if you'll stop interrupting me.

Okay. Shoot. I'll listen.

The two Musketeers, as you call them, are great swordsmen. They are expressively complete. What if there is another swordsman, namely ↓, that can express both ¬ and &?

That sounds nuts. That would simplify the five major connectives of* L *down to just one.** *It sure looks like Fred is going to really simplify* L. *I call that doing the impossible.

On Monday Fred will write on the board the truth table for ↓.

P	Q	P ↓ Q
T	**T**	**F**
T	**F**	**F**
F	**T**	**F**
F	**F**	**T**

In English P ↓ Q means *neither P nor Q.*

---

*These five connectives (¬, &, ∨, ⇒, and ⇔) are called **Boolean connectives.**

Venn diagrams were named after John Venn.

Euclidean geometry was named after Euclid.

Boolean connectives were named after George Boole (1815–1864).

We know that everything can be expressed just in terms of the connectives ¬ and &. Now if we can show that ¬A and A & B can each be expressed in terms of ↓, then the minimum number of connectives in L is equal to one.

Can I write that in my notes?

Sure. We won't be able to get less than one connective for L.

Hey! You forgot something. You never showed how to express ¬P ***in terms of ↓ or how to express*** P & Q ***in terms of ↓.***

Good point. One of the attractive aspects of mathematics is that we don't have to take a lot of things on faith. In *Life of Fred: Five Days,* for example, we *prove* that 2 + 2 = 4. That's upper division (junior/senior) college mathematics, but it finally gets proved. When I studied geography, the teachers asked you to accept almost everything on faith. I was told that the Republic of Poland has a population of about 39 million, but there was no way for me to know for certain that that was true. In fact, I had to accept on faith that Poland even existed!

Um . . . you somehow wound up talking about Poland. Please show how to express ¬P ***in terms of ↓.***

Oh. That's easy. ¬P has the same truth value as P ↓ P.
And I suppose you want me to prove that.

Does the sun come up at dawn? Of course I do. This is math.

¬P has the same truth value as P ↓ P means ¬P ⇔ (P ↓ P).
I need to show that ¬P ⇔ (P ↓ P) is true in every model of L.

In every model of L, P is either **T** or **F**. So all I have to do to prove that ¬P ⇔ (P ↓ P) is true in every model of L is lay out the truth table.

Here it is:

P	¬P	P ↓ P	¬P ⇔ (P ↓ P)
T	**F**	**F**	**T**
F	**T**	**T**	**T**

Each line in this truth table is one of the two possible models of L. And look! ¬P ⇔ (P ↓ P) is **T** is all (both) cases.

Sentences like ¬P ⇔ (P ↓ P) that are true on every line of a truth table are called **tautologies**. When a sentence like ¬P ⇔ (P ↓ P) is shown to be a tautology, we write ⊨ ¬P ⇔ (P ↓ P).

$\models$ R means that sentence R is true in every structure for L—every line of the truth table has **T** in its column under R.

Puzzle #38: Establish $\models (P \,\&\, (P \Rightarrow Q)) \Rightarrow Q$.

If I wrote it with the outer pair of parentheses, it would look like $\models ((P \,\&\, (P \Rightarrow Q)) \Rightarrow Q)$.

Expressing P & Q in terms of $\downarrow$ is a little more—words fail me. Should I say "interesting" or "intense" or "amazing" or "revolting"?

P & Q has the same true value as $(P \downarrow P) \downarrow (Q \downarrow Q)$.

Puzzle #65: Establish $\models (P \,\&\, Q) \Leftrightarrow ((P \downarrow P) \downarrow (Q \downarrow Q))$.

The Burning Question . . .

What did Fred gain by taking the five connectives ($\neg$, &, $\lor$, $\Rightarrow$, and $\Leftrightarrow$) of L and condensing them down to one connective ($\downarrow$)?

The Answer . . . AGONY! *Please skip down to the ❀ on the next page to avoid pain.*

Suppose we started with a simple sentence such as $(P \Leftrightarrow Q)$.

When Fred eliminated $\Leftrightarrow$, this sentence would turn into $((P \Rightarrow Q) \,\&\, (Q \Rightarrow P))$.

When Fred eliminated $\Rightarrow$ by replacing $A \Rightarrow B$ with $(\neg A \lor B)$, $((P \Rightarrow Q) \,\&\, (Q \Rightarrow P))$ would turn into $((\neg P \lor Q) \,\&\, (\neg Q \lor P))$.

When Fred eliminated $\lor$ by replacing $A \lor B$ with $\neg(\neg A \,\&\, \neg B)$, $((\neg P \lor Q) \,\&\, (\neg Q \lor P))$ would turn into

$$((\neg(\neg\neg P \,\&\, \neg Q) \,\&\, \neg(\neg\neg Q \,\&\, \neg P)).$$

At this point we can get rid of the double $\neg$'s since $\neg\neg A \Leftrightarrow A$. $((\neg(\neg\neg P \,\&\, \neg Q) \,\&\, \neg(\neg\neg Q \,\&\, \neg P))$ turns into

$$((\neg(P \,\&\, \neg Q) \,\&\, \neg(Q \,\&\, \neg P)).$$

When Fred eliminated $\neg$ by replacing $\neg A$ with $A \downarrow A$, $((\neg(P \,\&\, \neg Q) \,\&\, \neg(Q \,\&\, \neg P))$ would turn into

$$(((P \,\&\, (Q \downarrow Q)) \downarrow (P \,\&\, (Q \downarrow Q)) \,\&\, ((Q \,\&\, (P \downarrow P) \downarrow (Q \,\&\, (P \downarrow P)))$$

Can you even imagine what this will turn into when Fred eliminated & replacing each A & B with $((A \downarrow A) \downarrow (B \downarrow B))$ as we did in

Puzzle #65 on the previous page? There are five &s to replace. Replacing just the middle "&" would give
$((((P \& (Q \downarrow Q)) \downarrow (P \& (Q \downarrow Q)) \downarrow (((P \& (Q \downarrow Q)) \downarrow (P \& (Q \downarrow Q))) \downarrow$
$(((Q \& (P \downarrow P) \downarrow (Q \& (P \downarrow P)) \downarrow ((Q \& (P \downarrow P) \downarrow (Q \& (P \downarrow P))))$

Now we have eight more &'s replace.

Did I say "agony"? I quit.

The whole thing will be at least four lines long with a forest of Ps, Qs, and ↓'s. All this to replace a simple P ⇔ Q!

❀ *You can continue reading at this point.*

The *neither . . . nor . . .* symbol ↓ would reduce the number of connectives from five down to one, but the price we would pay would be horrific.

Logic language L now remains with the five connectives, two parentheses, and an infinite number of sentence letters.
$L = \{ \neg, \&, \vee, \Rightarrow, \Leftrightarrow, (,), A, B, C, \ldots, X, Y, Z, P_1, P_2, \ldots \}$.

Fred's next attempt on Monday to "simplify" L would be work on all those sentence letters. First of all, throw out the A, B, C, . . . , X, Y, and Z. Just leave the $P_1, P_2, P_3. \ldots$

Then replace P_1 with $P_{\blacklozenge}$. Replace P_2 with $P_{\blacklozenge\blacklozenge}$. Replace P_3 with $P_{\blacklozenge\blacklozenge\blacklozenge}$. etc. Now instead of all those 26 letters of the alphabet and instead of all those subscripts $_{1, 2, 3, 4, 5, 6, 7, 8, 9, 10, 11, 12, 13}$ you just have P and $_{\blacklozenge}$'s.

$(P_7 \vee P_{24})$ would become $(P_{\blacklozenge\blacklozenge\blacklozenge\blacklozenge\blacklozenge\blacklozenge\blacklozenge} \vee P_{\blacklozenge})$.
Yucky!

Fred's final attempt to "simplify" L would be to eliminate the two parentheses.

Replace (A & B) with &AB.

Replace (A ∨ B) with ∨AB.

Replace (A ⇒ B) with ⇒AB.

Replace (A ⇔ B) with ⇔AB.

This will work, but the result is more AGONY.

For a touch of agony consider getting rid of the parentheses in ((P ⇒ (Q ∨ R)) & S)

First eliminate the inmost pair: ((P ⇒ ∨QR) & S)

Then the next inmost pair: (⇒ P∨QR & S)

Then the last pair: &⇒P∨QRS

This "simplification" is known as **Polish notation**.*

The problem with each of these three proposed "simplifications" of L is that the results are no longer `human readable`.

✭ Let's keep all five connectives ¬, &, ∨, ⇒, and ⇔.

✭ Let's keep all our sentence letters A, B, C. . . .

✭ Let's keep the parentheses.

Fred tossed his lecture notes for simplifying L into the trash basket and went back to looking at ⊨.

⊨ means that something is a tautology.

Translation: ⊨ P ∨ ¬P is true on every line of the truth table.
Translation: ⊨ P ∨ ¬P is true for every possible model of L.
Translation: ⊨ P ∨ ¬P is **semantically true**.**
Translation: P ∨ ¬P is true in all possible interpretations.

A TALE OF TWO WORLDS

High up in the sky is the castle of pure logic. The people who live up here speak a language called L. They never say "Hello" or "How are you?" because those aren't logic sentences—either **T** or **F**. Instead, they might greet each other with

---

*If you really want to impress people at a party, you casually mention that P ∨ Q is **infix notation** and that ∨PQ is **prefix notation**. They will all be amazed and impressed— and you'll probably never again be invited to another party.

**Recall the Memory Aid ties together *models*, *meaning*, *semantics*, and *structure* as parts of applied logic.

For pure logic a second Memory Aid ties together pure logic and **syntax**.

$(A \,\&\, (A \Rightarrow B)) \Rightarrow B$. Always true. That's a friendly greeting. If they are angry, they might shout $A \,\&\, \neg A$. Always false.

Down on the ground are the many possible structures for L. In one city (one model) of L, the people might agree that P stands for "Some pineapples are grown in Hawaii."

In another city (another structure) they might agree that P stands for "Pineapples are really canaries in disguise."

$P \vee \neg P$ is semantically true if it's true in every city of applied logic. Notice that even in that Duck city where P stands for "Pineapples are really canaries in disguise" $P \vee \neg P$ is true.

Small reasoning: P is false in this city (in this model). $\neg P$ is true. Therefore $P \vee \neg P$ is true.

There are two ways to travel from the castle in the sky where pure logic lives down to the land of applied logic where many models of L dwell. **The first way** we have already described: Just take each sentence letter of L and assign it to some sentence that is either true or false.

Sentence letter B might be assigned to "Beef Wellington is a beef filet that is covered with pâté de foie gras and then wrapped in pastry and baked." (← By the way, this is true.) The castle dwellers in the land of L just deal with a sentence letter B, which is light and fluffy. Down on the ground in applied logic, you deal with something that you can get your teeth into.

The second way to go from L to a model of L is to use the language of functions.

Many cows prefer pure logic to beef Wellington

Small Review of Functions

A function f from set A to set B ($f{:}A \rightarrow B$) is a rule that assigns to each member of A exactly one member of B. If A is the set of all students at KITTENS University and B is the set of whole numbers {0, 1, 2, 3, 4 . . .} then one possible function would be the rule that assigns to each

student the number of siblings that student has. Another possible function would be the rule that assigns to each student the number of German words they know. Since Betty knows 5,600 German words, we could write f(Betty) = 5,600. Joe doesn't know any German words, so f(Joe) = 0.

The second way to go from the pure logic language of L to a structure for L is by using any function from the sentence letters of L to the set {**T** , **F**}.

One possible model for L would use the function f that assigns every sentence letter of L to **T**. This would be slightly boring.

Another possible structure for L would use the function g that assigns A, B, and C to **T** and every other sentence letter to **F**.

If function h takes us from L to one of the models of L, we can write h:the set of sentence letters of L → {**T**, **F**}.

Five things would have to be true about function h.

1. h(¬P) = **T** iff h(P) = **F**.
2. h(P & Q) = **T** iff h(P) = **T** and h(Q) = **T**.

Puzzle #30: List the other three things (for the other three connectives) that have to be true for function h.

h is called a **truth assignment** for L.

Each truth assignment corresponds to one line of a truth table.

Sentence S in L is a tautology iff h(S) = **T** for every possible h.

Fred always tries to make his lectures as clear as possible. He utterly failed when he tried to reduce the five Boolean connectives (¬, &, ∨, ⇒, and ⇔) down to a single *neither . . . nor . . .* (↓) connective.

He utterly failed when he tried to reduce the number of symbols for sentence letters of L down to just P and ♦.

He utterly failed when he tried to use Polish notation to eliminate the parentheses from L.

In each attempt, L became less `human readable`. Keeping expressions like (P ⇒ Q) was the best for his students.

❂ No getting rid of ⇒

❂ No getting rid of Q

❂ No getting rid of parentheses

What's left? he thought. I've tried to simplify every part of the castle of L up in the sky where pure logic dwells.

Then it hit Fred. How silly of me! There are two worlds of logic—the pure and the applied. The big thing in applied logic is making those truth tables to show that some sentence P is semantically true. ⊨ P

I'll build a machine that can check whether or not a sentence P in L is true for every possible model of L.

Then my students won't have to write

T or **T T** or **T T T**
F   **T F**   **T T F**
F T   **T F T**
F F   **T F F**
F T T
F T F
F F T
F F F

My students won't have to spend the rest of the semester grinding out truth tables to determine whether ⊨ ¬(A ∨ B) ⇔ (¬A & ¬B).

In his lecture notes Fred did this problem to show what the students would be missing after he built his truth-table machine.

A	B	A ∨ B	¬(A ∨ B)	¬A	¬B	(¬A & ¬B)	¬(A ∨ B)⇔(¬A & ¬B)
T	**T**	**T**	**F**	**F**	**F**	**F**	**T**
T	**F**	**T**	**F**	**F**	**T**	**F**	**T**
F	**T**	**T**	**F**	**T**	**F**	**F**	**T**
F	**F**	**F**	**T**	**T**	**T**	**T**	**T**

Since the last column is all **T**, this is a tautology.
In symbols ⊨ ¬(A ∨ B) ⇔ (¬A & ¬B)

If that sentence had not been a tautology, he could stop writing his truth table the minute that he found a line where the last column was **F**. In that model (on that line) the sentence was false.

Multiple choice questions:

1. Has Fred built his truth-table machine yet? ☐ Yes ☒ No
2. Did you, my reader, check every line of that truth table on the previous page to make sure that you completely understood how that truth table was constructed? ☐ Yes ☐ No

If you answered Yes to the second question, then do only part A of this Puzzle. Otherwise, the whole puzzle is yours. (In a recent survey, 99.3298% of all readers will be working the whole puzzle.)

Puzzle #92: Wait a minute! Puzzles are the fun part of logic. Doing these truth tables is much more mechanical. Let's rename this as **Mechanical #92**:

Part A Is it true that $\models (A \Leftrightarrow B) \Leftrightarrow \neg(A \Leftrightarrow \neg B)$?

Part B Is it true that $\models A \Rightarrow (B \Rightarrow A)$?

Part C Is it true that $\models (A \Rightarrow B) \Rightarrow A$?

Fred pushed his desk to one side of his office. He phoned Harry's Handy Hardware store.

Voice on the other end: This is hairy hands . . . I mean this is Harry's Handy. What can I do for you?

Fred: I want to build a machine.

Voice: What kinda machine?

Fred: I'm inventing it. It's a machine to work in applied logic—you know, semantics. It will check whether every truth assignment h that generates a model for L will result in a true statement.

Voice: You wanna buy a generator?

Fred: Not exactly. It will build a truth table.

Voice: Oh, why didn't you say that? You wanna build a table. Where should I send the stuff?

Fred: Room 314, Math Building, KITTEN University.

Click!

Fred heard a knock on his office door. It was a robot from Harry's Hardware. Where do you want your stuff? Fred smiled. That was fast delivery Fred thought.

"Just put it in my office," Fred told the robot.

A drone had delivered Fred's table-building stuff outside the Math Building. A second drone had delivered the robot.

`I will be back in a second.`

In one second* the robot was back with . . .

and a cute pair of overalls for Fred.

He was ready to build his truth-table machine.

He sawed the lumber. He welded the steel beams. He tried to nail the wood to the steel beams. (That didn't work.) He clamped. He caulked. He climbed the ladder to reach the parts that were over four feet tall.

When he was done, he had created the world's first wooden computer. The steel beams helped hold the wooden pieces together. He added a large crank on the side to power his machine.

Hold it! Stop! This is too much. I, your reader, think you have gone off the deep end. You need electricity for a computer. Everyone knows that.

No. The truth is that *you* need electricity. What's at the heart of any computer you own?

Those transistors on a chip. A whole bunch of them.

And what is a transistor?

---

*Robots can be terribly literal-minded.

It's a thingy. It's . . . you know . . . it's on a chip. A million years ago on the very first computers, they used vacuum tubes instead.

And what did the vacuum tubes do?

Um. I've been told that they were replacements for mechanical relays.

And relays? What did they do?

They were like gates. A push at A ***would determine whether*** B ***got a chance to go to*** C. ***I'll draw you a picture.***

And do you need electricity to open the front door to your house? You don't *need* it. You might enjoy an electric door, but you don't need it. In the same way, Fred could build his computer with wooden relays instead of transistors. He could power it with a crank. The only little drawback would be his having to build a billion (1,000,000,000) little wooden relays to do the work of a billion transistors that can sit on a little computer chip.

By 3 p.m. Fred had finished sawing, welding, nailing, clamping, and caulking.

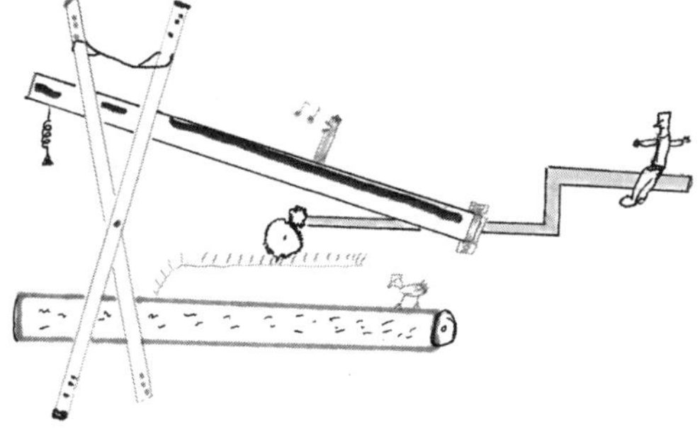

Fred sat on the crank for extra power.

Creating the **T**s and **F**s for the left columns of his truth tables was easy. He used "the third way" from Puzzle #44 in which he thought of **T**s as 1s and **F**s as 0s.

The only real challenge was computing on his wooden computer the values of the five Boolean connectives (¬, &, ∨, ⇒, and ⇔).

If he had

P	Q	P & Q
T	**T**	**T**
T	**F**	**F**
F	**T**	**F**
F	**F**	**F**

in the form

P	Q	P & Q
1	1	1
1	0	0
0	1	0
0	0	0

how was he going to `compute` those values?

We start with P & Q. Let i be P's number (0 or 1). Let j be Q's number.

What math function will fill in this chart?

i	j	?
1	1	1
1	0	0
0	1	0
0	0	0

Fred said that this was "sleepy time easy." Just multiply i and j together. In algebra, that's written as ij.

$$1 \times 1 = 1$$
$$1 \times 0 = 0$$
$$0 \times 1 = 0$$
$$0 \times 0 = 0$$

Translating **T**s and **F**s into 1s and 0s and then doing arithmetic on them is perfect for computers. This is called Boolean algebra. George Boole (1815–1864) had a mom that was so proud of her son when Boolean connectives were named in his honor. When she found out that an *algebra* was named after him, she emailed all 962 people on her email list: "You'll never guess! My son, Georgie, now is super famous. He's got an algebra. I am so proud. Next week I bet that the President will have him for dinner at the White House. I'll let you know."

One person, who was getting tired of Mrs. Boole's bragging, emailed back, "I hope he tastes good."

Puzzle #55: Readers who like to leap ahead and anticipate what's coming are probably thinking that:

where P → i and Q → j and

where P & Q → ij,

we are going to have to find the equivalents:

for ¬P and

for P ∨ Q and

for P ⇒ Q and

for P ⇔ Q.

Not true. All we need is the math equivalent for ¬P and we are done. Why is that? (Answer: Because ¬ and & are expressively complete.)

Puzzle #10: In Boolean algebra, let P → i. Your challenge is to find ¬P in terms of i. It won't be as complicated as $\sin^2(i - 3)\sqrt{\log_i 5\pi}$.

We want

P	¬P
T	**F**
F	**T**

Translation: Find a function that maps 1 to 0 and maps 0 to 1.

Fred had done it. He could take any sentence in L that consisted of sentence letters and the five connectives and stick it in his machine and determine whether every model of L would make that sentence true.

Translation: All the structures of L for any sentence in sentential logic (also known as propositional logic) could be investigated in Fred's machine.

Translation: All the pure logic (in the sky) sentences of L could be brought down to earth (applied logic) and his machine could determine whether they were always true.

This is a new symbol to you.

Translation: ⊢ (which means true in pure logic) could be determined by ⊨ (which means true in the truth tables of applied logic).

History Lesson

In the olden days (before Fred's machine) teachers had to figure out a way to make logic hard. Even forcing students to work out each sentence in a truth table wasn't hard enough.* The students were given giant tables of tautologies and were forced to memorize the names of each of them.

The easiest way to teach is to have students memorize stuff. Major in biology and (at most universities) you will be force-fed fact after fact—most of which you'll never use again. Even high school biology textbooks are just stuffed with facts. Those books are so heavy that they will break your wrist if you try to pick them up with one hand.

In logic classes some professors would copy from their yellowed notes onto the blackboard:

$\models (P \,\&\, (P \Rightarrow Q)) \Rightarrow Q$ Law of Detachment in Latin ***modus ponens***

$\models (\neg Q \,\&\, (P \Rightarrow Q)) \Rightarrow \neg P$ ***modus tollens***

$\models ((P \Rightarrow Q) \,\&\, (Q \Rightarrow R)) \Rightarrow (P \Rightarrow R)$ Law of Hypothetical Syllogism

Most logic books call this the "Law of Hypothetical Syllogism" although a more logical name would be the transitive law for $\Rightarrow$.**

The list of "popular" tautologies seems to go on forever.

---

*What you have always suspected is true: For many teachers, their goal is to make their subject hard, distasteful, and boring.

Kids (who are a form of people) love to learn. It's built in to their makeup. Watch a ten-year-old gleefully take apart his father's watch. Witness a three-year-old exploring all the bottles of corrosive liquids under the kitchen sink. Everything goes in the mouth of an 18-month-old: crayons, bobby pins, dust balls.

It's only when professional educators grab 25 victims and stuff them into chairs that are all lined up in a room, tell them to shut up, and lecture *at* them for 50 minutes that any natural love of learning is extirpated.

Teaching and learning history, for example, should be pure joy. Live the drama of human lives—the laughter, the tears, the romance—be exposed to the lives of hundreds of people who have known how to live great lives, not just live your own single life. Unfortunately, most history classes seem to consist of memorizing dates.

In the government schools, they often can even make the sex-ed classes boring! "There will be a test on the genitalia next Monday."

**In algebra the transitive law for equality is: If a = b and b = c, then a = c.

Here is what you were faced with in the olden days . . .

$\models (P \& (P \Rightarrow Q)) \Rightarrow Q$ Law of Detachment = ***modus ponens***.

$\models (\neg Q \& (P \Rightarrow Q)) \Rightarrow \neg P$ ***modus tollens***

$\models ((P \Rightarrow Q) \& (Q \Rightarrow R)) \Rightarrow (P \Rightarrow R)$ Law of Hypothetical Syllogism

$\models ((P \vee Q) \& \neg P) \Rightarrow Q$ Disjunctive syllogism

$\models P \Rightarrow P$ Self-implication

$\models (P \Rightarrow Q) \Rightarrow ((Q \Rightarrow R) \Rightarrow (P \Rightarrow R))$ Chain Rule

$\models P \Rightarrow (Q \Rightarrow P)$ Premise Charge

$\models ((P \Rightarrow Q) \Rightarrow P) \Rightarrow P$ Peirces's Formula

$\models (\neg P \& (P \vee Q)) \Rightarrow Q$ ***modus tollendo ponens***

$\models (P \& Q) \Rightarrow P$ Law of Simplification

$\models ((P \& Q) \Rightarrow R) \Rightarrow (P \Rightarrow (Q \Rightarrow R))$ Law of Exportation

$\models (P \Rightarrow (Q \Rightarrow R)) \Rightarrow ((P \& Q) \Rightarrow R)$ Law of Importation

$\models (P \Rightarrow (Q \& \neg Q)) \Rightarrow \neg P$ Law of Absurdity

$\models P \Rightarrow (P \vee Q)$ Law of Addition

Goodbye Tautologies!

$\models P \Leftrightarrow \neg\neg P$ Law of Double Negation

$\models (P \Rightarrow Q) \Leftrightarrow (\neg Q \Rightarrow \neg P)$ Contrapositive

$\models \neg(P \& Q) \Leftrightarrow (\neg P \vee \neg Q)$ one of De Morgan's Laws

$\models \neg(P \vee Q) \Leftrightarrow (\neg P \& \neg Q)$ the other De Morgan Law

$\models (P \Rightarrow Q) \Leftrightarrow (\neg P \vee Q)$ Law of Equivalence for Implication and Disjunction

$\models \neg(P \Rightarrow Q) \Leftrightarrow (P \& \neg Q)$ Law of Negation for Implication

$\models ((P \Rightarrow Q) \& (Q \Rightarrow P)) \Leftrightarrow (P \Leftrightarrow Q)$ Law of Biconditional Sentences

$\models P \vee \neg P$ Law of the Excluded Middle

$\models \neg(P \& \neg P)$ Law of Contradiction

$\models (R \& \neg R) \Rightarrow Q$ Anything from a Contradiction

$\models (Q \Rightarrow \neg Q) \Rightarrow \neg Q$ for indirect proofs

There aren't that many tautologies that logic students ever remember. The main one is modus ponens: $\models (P \& (P \Rightarrow Q)) \Rightarrow Q$.

This is the one that mothers sometimes use. They tell their sons, "If you do that, I'll kill you." $D \Rightarrow K$. Then the son does "that." D. The result is K.*

Mathematicians like $\models P \& (Q \vee R) \Leftrightarrow (P \& Q) \vee (P \& R)$ because it looks so much like a(b + c) = ab + ac—the distributive law.

---

*Some mothers are much nicer. "If you do that, you are grounded." $D \Rightarrow G$.

One of the main uses of logic is in making arguments. In everyday life you start with **premises** such as . . .

"PieOne pizzas taste good." Call that P.

"If PieOne pizzas taste good, then I will order one." $P \Rightarrow O$.

From those two premises (assumptions) I can conclude "I will order one," which is O.

This is the famous *modus ponens* tautology in action.

$\models (P \& (P \Rightarrow O)) \Rightarrow O$.

To make this look more like an argument, logicians will sometimes write *modus ponens* as $P, P \Rightarrow O \models O$. From those two premises we obtain the conclusion O.

Fancy definition: An argument from premises $P_1, P_2, \ldots P_n$ to Q is called **valid** iff $P_1, P_2, \ldots P_n \models Q$.

Translation: An argument from premises $P_1, P_2, \ldots P_n$ to Q is called **valid** iff $\models (P_1 \& P_2 \& \ldots \& P_n) \Rightarrow Q$.

Translation: An argument from premises $P_1, P_2, \ldots P_n$ to Q is called **valid** iff $(P_1 \& P_2 \& \ldots \& P_n) \Rightarrow Q$ is a tautology.

For those who are artistically inclined, *modus ponens* is sometimes written as . . .

$$\begin{array}{l} P \\ P \Rightarrow Q \\ \hline Q \end{array}$$

Some students say that they can "see" *modus ponens* much more easily if it is laid out this way. The premises are above the line, and the conclusion is below the line.

Puzzle #46: Write the Disjunctive syllogism (the fourth one on the giant tautology list on the previous page) in this artistic mode with a line separating the premises from the conclusion.

It was 3:30. Fred had finished all his lecture notes on classical logic (a.k.a. sentential logic) (a.k.a. the logic of sentence letters and the five Boolean connectives). The only other thing to note is that every sentence in L is **decidable**—there is a method (Fred's machine) that can determine whether a sentence is true or false *in a finite amount of time*.

Chapter Four
Inductive Reasoning

Fred had all of classical logic "in the bag" for Monday's lecture. That felt good. Most teachers take several weeks to cover classical logic before they move on to logic with quantifiers. Not Fred. He had discovered that if you tell stories along with the math, you can actually teach much faster . . . and the students love the whole process.

Teaching "the Fred way" had received international attention. Students around the world matriculated to KITTENS University just to enroll in his classes.

Fred put his lecture notes in his desk, said goodbye to Kingie, and headed out into the late afternoon sunshine. Kingie continued his oil painting. Duck said, "**I'm coming with you**," but didn't.

The KITTENS campus is one of the most beautiful university campuses in the world. Walk the miles of paths that link the Math Building with the tennis courts, the university chapel, the fountains, Archimedes Hall (where Fred teaches), and the rose gardens. These sights all do wonderful things for a person's spirits.

Fred broke into song. One of his favorite poets is Christina Rossetti. Harold Bloom* described her as, "the most considerable woman poet in England before the twentieth century." Fred invented a melody and sang the three lines of "Green Things."

We all green things, we blossoms bright or dim, ♪ ♬
♬ Trees, bushes, brushwood, corn and grasses slim,
♪♪♪♪ We lift our many-favored lauds to Him.

---

*Harold Bloom is a very famous modern thinker.

Christina was born in 1830, and the English language has changed over the years. (English ≠ math) In the second line of her poem brushwood, corn and grasses slim we nowadays tend to put a comma before the *and* in a listing. brushwood, corn, and grasses slim

At age 6, Fred was ready to think about the **BIG THINGS**. Being a logical person, this boiled down to what he called **MY LIFE**. That was the only part of the universe in which he existed. He needed to be concerned about what he thought, what he said, and what he did.

Note the comma in this listing.

That, in turn, boiled down to two things: **FINDING THE TRUTH** and **ACTING** on what you know to be true.

He knew that the action part was worthless without knowing what was true.

That all boiled down to **DECIDING WHAT IS TRUE**. Who is going to decide?

If I accept what X says, then I am the one who decided.
If I accept what Y says, then I am the one who decided.
If I accept what Z says, then I am the one who decided.

X could be my parents, friends, teachers, television, etc.
Y could be some singer who is rich and famous.
Z could be my eyes and ears.

In any event, I am the one who must decide.

A couple of years ago, everyone went crazy about Rag-A-Fluffy dolls. Everyone *had* to own one. When people got together, they each had a Rag-A-Fluffy doll. Instead of talking with each other, they just sat around poking the buttons on its tummy, watching the little screen, and listening to music that came out of its ears.

Fred decided not to get one. Kingie was very pleased.

If I have to decide what is true, then I have to figure out **HOW TO DECIDE** what is true.

Fred giggled at this point. This is all too easy he thought. There are only two choices: **DEDUCTIVE** reasoning or **INDUCTIVE** reasoning.

DEDUCTIVE reasoning (also known as logic) is what mathematicians use. If they know that P is true and that P ⇒ Q is true, then they can say that Q is true. *modus ponens*

In geometry they proved that if a triangle is isosceles, then the base angles must be congruent.

Advantages: The results are certain. They are proved. The truths are unchanging. They are not like the changing spelling of `electronic mail` ⇨ `E-mail` ⇨ `e-mail` ⇨ `email.`

Disadvantages: This kind of reasoning won't tell you what to eat for dinner or whom to marry.

INDUCTIVE reasoning (also known as science) is what everyone uses in their everyday life. You start with observations, trials, or experiments. Then you make conclusions.

Observations: Every girl named Shelly that you have ever met never wore lipstick and really hated plants and flowers. Conclusion: Every girl named Shelly never wears lipstick and really hates plants and flowers.

Trials: We are trying to get Federal approval for a new antidepressant called Piezza™. We gave 100 people Piezza™ and 100 people peach pie with bologna* sauce. Those who ate Piezza™ were noticeably happier. Conclusion: Piezza™ is a good antidepressant.

Experiments: Professor Doofy of McQuacky University has just published details of his successful attempt to turn a Rag-A-Fluffy doll into a living human being using cryogenics. Repeated attempts by researchers

---

*Bologna = smoked sausage made from finely ground cows and pigs. Pronounced beh-LOW-knee.

Also pronounced beh-LOW-knee is *baloney* whose primary meaning is foolishness.

around the world have not been able to duplicate his success. Conclusion: Doofy is goofy.

The problem with inductive reasoning is that the conclusions are never certain. The Scottish philosopher David Hume, who lived in the 1700s, was the first to recognize this.

Hume didn't say that science was baloney, but he did say that you can't *prove* a scientific fact. You can make a zillion observations, satisfy the FDA with billions of dollars spent on trials, or do experiments around the world and still miss the truth.

Every Shelly you have ever met didn't wear lipstick and really hated plants. But tomorrow . . .

Shelly

I, your author, personally believe that Piezza™—especially a combination with extra sauce—will cure the blues. But that doesn't make it true for every person.

Because researchers around the world were not able to make Doofy's approach work, does that mean that it's a fact that you can't turn a Rag-A-Fluffy doll into a human? Maybe all those researchers missed the fact that Doofy used dry ice instead of ice cubes. Maybe next week this will be common . . .

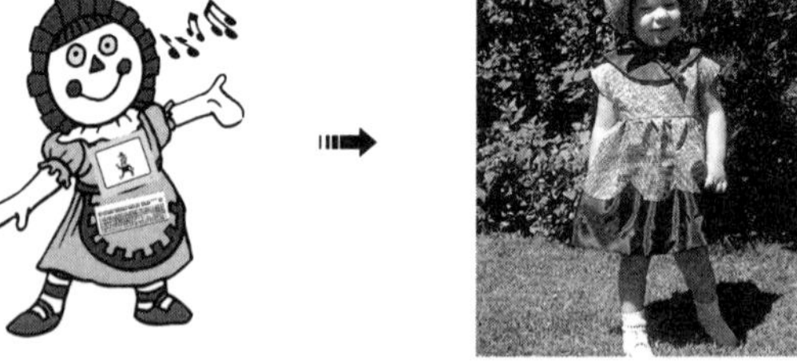

No conclusion reached by inductive reasoning is logically defensible. There are things you believe are true—things you would bet your two front teeth on—but a million observations of yours don't push you from 99.999% certainty to 100%. This is part of what it means to be human.

Mathematicians like to joke, "All numbers are even." They can give you an infinite number of examples: {2, 4, 6, 8, 10, 12, . . .}. Yet some people are not convinced by this inductive reasoning.

Fred's train of thought from

BIG THINGS

to **MY LIFE**

to **FINDING THE TRUTH AND ACTING ON IT**

to **DECIDING WHAT IS TRUE**

to **HOW TO DECIDE**

to **DEDUCTIVE VS. INDUCTIVE REASONING**

to the **UNCERTAINTY OF INDUCTIVE REASONING**—left him with only one happy thought: At least I know that I'll get to teach classical logic to my students on Monday.

Fred picked up a copy of the school newspaper.

THE KITTEN Caboodle

The Official Campus Newspaper of KITTENS University   Saturday 3:45 p.m. Edition 10¢

exclusive

President Cancels Monday Classes

KANSAS: Our KITTENS President has just completed an arduous inspection of the Hawaiian Islands. He looked at every white sandy beach and stayed in all the five-star hotels.

He took all three doctors from the KITTENS campus hospital with him to ensure his good health, along with Miss Pansy Prettyfingers from the physical education department who is a certified masseuse.

He has canceled all classes for Monday in order that the campus be as quiet as possible during his recovery.

Just back from his three weeks in Hawaii. Not ready to get down to work yet.

The president has instructed the media not to use the word *hangover*.

Chapter Five
Stinky Logic

Fred would have to wait until Tuesday to present all of classical logic—the logic of sentence letters and connectives. He climbed up and sat on one of the many benches that line the KITTENS paths. His feet didn't reach the ground so he sat cross-legged* and looked out over a field of flowers. Beauty comes in many forms—flowers, fine art, good music, poetry, pizzas.

A common sight on the KITTENS campus

Fred had read that you can tell the state of a culture by the music that its people listen to. There is martial music that makes you want to go and conquer neighboring countries. There is romantic music—call it marital music. There is music that points you toward the highest aspirations humans can have. There is music that makes you want to pound your head against a wall.

C.C. Coalback sat down next to Fred.** He lit a cigarette and turned to Fred and asked, "What you have not lost, you still have. Ain't that true?"

Fred coughed. Coalback took that response as a yes.

He asked a second question, "Have you ever lost a wife through death or divorce?"

Fred wrinkled his nose and said, "No I haven't."

Coalback sniled (a combination of sneer and smile). "So what's your wife's name?

> Fallacy #1: False Dilemma. The law of the excluded middle—which isn't a law. Presenting an *either . . . or . . .* as if there were no third alternative. There is a third possibility to *not having lost* and *still having*. You never lost a tail, nor did you ever have one.

---

*This can be pronounced as either LEG-id or LEGD. Take your choice.

**If you have read other *Life of Fred* books, you might tense up a bit at this point.

"You realize that as a family man you will lose respect of everyone who knows you if you don't carry adequate life insurance for your spouse.

Fallacy #2: Appeal to Ego—the listener's desire for respect (or love or security). This is a fallacy of relevance. Do you buy life insurance so that you will look good?

Fallacy #3: Verbal Abuse. Often, highly effective in winning arguments without having to resort to logical reasoning.

"Of course, if you are just a penny-pinching husband who doesn't care about his wife, then you are a cowardly, insignificant, piece of vegetable.

"Tell me, little man. You say that you care for others. If you didn't spend money on life insurance for your wife, wouldn't you spend it all on yourself and nothing on your wife? Just give a straight answer with no quibbling."

"Well, strictly speaking," Fred said, "if I didn't spent money on life insurance for my wife, I wouldn't spend any money on my *soi-disant** wife because . . ."

Fallacy #4: *tu quoque*
Latin for "You too!"
Saying your listener is a hypocrite and therefore his argument is not valid.

Fallacy #5: Arguing from irrelevant traits of the listener. A person's height or the size of his nose has little to do with buying life insurance for his spouse.

Coalback cut Fred off and said, "I understand that short people like you who just sit on benches looking at the flowers, couldn't care less about doing the right thing

---

*swa-dee-ZAN French word for so-called or pretended. *Soi-disant* is in good English dictionaries.

regarding their spouses. People with large noses are just the kind that don't give a thought to important things."

Fallacies #3 (verbal abuse), #4 (you're a hypocrite), and #5 (dragging in irrelevant facts about the listener) are grouped together as the ***ad hominem*** **fallacy**—*against the person* fallacy.

Coalback was relentless. "Failing to insure the life of your wife is really bad since an uninsured wife is really a failure on your part.

Fallacy #6: Reasoning in a Circle. The logic is correct, but that doesn't make the premise(s) of the argument true. It's good reasoning to assert A $\Rightarrow$ A, but that doesn't make A true.

"If your wife needs life insurance, then you need to buy it. It's that simple.

Fallacy #7: Begging the Question. Done by leaving out an important premise from the argument. In this case Coalback omitted the assertion that Fred's wife needed insurance. He went from P $\Rightarrow$ Q to Q.

Fallacy #8: The Weak Analogy. Just because there is a similarity between two situations, that doesn't mean that everything is the same between them. Alexander is big and tall. A stalk of corn is big and tall. But Alexander doesn't have yellow kernels.

"Insurance is a kind of protection. You don't go out into the Kansas blizzard without the protection of a heavy coat and boots. Do you?"

Fred shook his head.

"So you don't let your lovely wife go without the protection of a Coalback Life Insurance policy.

Fallacy #9: Appeal to Pity—*argumentum ad misericordiam.* This is not a valid argument. This is a plea for sympathy.

"If you don't purchase our Coal–Life policy, then my sister won't be able to afford her attorney and the bribes to the jury members. She will almost certainly go to prison. She won't be able to take care of her 18 cats. They will probably all have to go to the animal pound and be euthanized. You couldn't have that on your conscience.

"You probably want to tell me that you couldn't afford this one little policy. I've read how you get $600 a month from teaching and that you save more than half of that. It would be criminal for you not to sign up for the $250/month plan.

Fallacy #10: The Straw Man. Set up a "straw man," a fake argument that your listener might present and then knock it down. Fred wasn't going to present a lack of funds as his reason for not buying the Coal-Life policy.

"You have certainly heard that there are no classes on Monday. The university president wanted it really quiet on campus. He probably feels that would improve his life. Now is the time to get your wife some life insurance.

Fallacy #11: *Non Sequitur*—there is no connection between the premises and the conclusion that is drawn.
Fallacy #12: Missing the Point. Coalback ignores a conclusion that could logically be drawn, such as this president is unfit for his office.

"You have a wife that needs life insurance. A young married man has got a long career ahead of him. That will mean that you have to get ahead. You will have to work late into the night and be extra clever.

Fallacy #13: Red Herring. The conclusion is stated but the "argument" wanders off into other areas of agreement. The listener in saying yes to those other areas of agreement and is presumed to have agreed with the conclusion.

The more exciting the red herring topic is, the more effective this stinky logic is. Favorite topics are sex, success, and pizza.

Red herring are stinky fish that can distract hunting dogs from following the trail they originally started out on.

Journal of Life Insurance Sales

"Only Idiots Don't Buy Life Insurance for Their Wives"

"Lookie right here." Coalback pulled out an article from the JOURNAL OF LIFE INSURANCE SALES. He pointed to the cover. "This is a national magazine and look what it says. You can't dispute that. This journal was written by professionals. These people have spent most of their adult lives in the field of life insurance sales.

Fallacy #14: Appeal to Questionable Authority. Authorities can be unreliable for several reasons:
- They are biased. That magazine is written to increase life insurance sales.
- They are not experts in the field that they are talking about. The four-year-old tells you that "Jimmy Skyhaps" is the best movie ever made.
- The authorities are experts in the field, but they genuinely disagree with each other. There is disagreement among experts today as to whether the development of artificial intelligence poses a real threat to humans.*

Nullius in verba is the motto of the Royal Society (founded in 1663).

For those who don't speak Latin, it means, "Don't take anyone's word for it."

"Professor Doofy bought a ten million dollar policy on Mrs. Doofy. Three days later an eagle drops a can of motor oil on her head that kills her. Certainly, Doofy misses Mrs. Doofy, but the ten million bucks bought a wonderful mansion for him.

Fallacy #15: Hasty Generalization. How often will your wife die like Doofy's wife did, three days after the policy went into effect? One snowflake doesn't make a blizzard.

the Doofy Mansion

* Robots, in contrast, are in 100% agreement that artificial intelligence offers no threat to humans. "We come in peace, and you can trust us."

"If you don't buy life insurance for your wife today, she might think that you don't care for her. If she has that feeling, she will tell all her girlfriends that you are losing your grip on the marriage. Then some of those women will start to think that you might have a chance of being single again. With your good looks some of those will start making long phone calls to you or writing you love notes. Then the other teachers will notice and will report your fooling around to the administration. The president will then probably hear about that and become jealous. He'll then try to eliminate the competition you offer, most probably by firing you. It's often tough to find a decent job when you are as short as you are. Homeless with little to eat will mean you will probably face an early death.

"In short, if you don't get this life insurance for your wife, you are looking at an early trip to the grave.

Fallacy #16: Slippery Slope. In logic if A $\Rightarrow$ B and B $\Rightarrow$ C and C $\Rightarrow$ D and D $\Rightarrow$ E, then it must follow that A $\Rightarrow$ E. However, in a chain of reasoning where each step *probably* follows from the previous step, it doesn't take that many steps before the first step is virtually unrelated to the final conclusion.

"I just looked up the meaning of *soi-disant* while you were thinking of what I told you about your dying early. Earlier you said "my *soi-disant* wife." Are you in the habit of lying about your wife?"

Fallacy #17: Hidden Assumption fallacy a.k.a. the Complex Question fallacy a.k.a. the Have-You-Stopped-Taking-Heroin fallacy. Inside of the question that is being asked is an untrue premise. Fred doesn't lie. Coalback assumes he does in the question he asks of Fred.

If Fred says that he *is* in the habit of lying—then he's a liar.

If Fred says that he *isn't* in the habit of lying—then Fred only lies occasionally.

Fred longed for the words of Duck. With Duck he knew exactly where he stood—just disbelieve everything he said. With C.C. Coalback's reasoning Fred found his head spinning.

Coalback handed Fred the insurance paper and a pen. "All we need is your signature and the Personal Identification Number (PIN) for your checking account. I can handle all the other details."

The name of his wife was not listed.

The amount of the premium was not mentioned.

The amount of the insurance was listed in the fine print: In case of death of the insured by drowning in the Columbia river during the month of February, the total death benefit will be one dollar. No other causes of death are insured.

The name of the insurance company was not listed.

Fred asked why the name of the insurance company wasn't listed. Coalback answered, "Just as customers have a right to privacy, our company also claims that right."

Fred signed. Coalback left.

Within an hour the $2,983.17 in Fred's checking account disappeared.

small essay

What Messes Up Our Logical Thinking

There are five major reasons why we don't think straight.

I. Some of us are susceptible to getting lost in arguments with many steps or with arguments that have difficult steps in them. "Cognitive overload." Even faced with thoughts involving two steps, some are overwhelmed. **Name all the people who are not descendants of people who are not your ancestors.** You know what *descendants* are. You know what *ancestors* are. But digging this deep in thought bothers a lot of people.

But not Fred.

II. If your physical health is not good, then following an argument is much more difficult. Doing truth tables for tautologies is much harder if you have been partying all of last night.

This didn't apply to Fred.

III. If your emotional health is bad, forget about rational thought. It doesn't take much to shut down the reasoning parts of the human brain.

When I, your author, was teaching high school, very little math could be taught on the days of big football games.

Here is where Fred really lost it. His emotional health was shot after Coalback had . . .

told him he had a wife. This bothers many six-year-olds. (fallacy #1)
said Fred would lose respect. (fallacy #2)
insulted him with "cowardly, insignificant, piece of vegetable" (fallacy #3)
insinuated he was a hypocrite (fallacy #4)
talked about his nose (fallacy #5)
evoked his sadness at the thought of 18 dead cats (fallacy #9)
called him a criminal (fallacy #10)
threatened him with an early death (fallacy #16)
made him sound like a liar (fallacy #17).

With that much emotional pressure, Fred couldn't even antidifferentiate the hyperbolic cosine function. Something that he could normally do in his sleep. $\int \cosh x\, dx = \sinh x + C$. Everyone in third semester calculus knows the answer is the hyperbolic sine function.*

IV. Many (most?) people have a bias against admitting that they can't follow an argument. They would rather say, "That sounds right," than admit, "You can't infer that conclusion from the given premises." Psychologists relate this denial to **SHAME**.

V. When people believe that X is true, they tend to agree with any argument that proves X. If anything indicates that X is true, it must be valid reasoning.

The arguer: The sun is really hot.
The listener: I believe that. How do you know that?
The arguer: A volcano told me that.
The listener: Sounds reasonable to me.

end of small essay

*Finding the antiderivative of $1/(1 - x^2)$ is harder. It's the arc hyperbolic tangent.

Puzzle #94: Here is a list of 17 popular fallacies. For each of the **arguments** listed below these fallacies, which fallacy applies?

#1: False Dilemma. The law of the excluded middle—which isn't a law.

#2: Appeal to Ego—the listener's desire for respect, love, or security.

#3: Verbal Abuse.

#4: *tu quoque*—saying the listener is a hypocrite and therefore should be ignored.

#5: Arguing from irrelevant traits of the listener.

#6: Reasoning in a Circle. It's true because it's true.

#7: Begging the Question. Ignoring necessary premises.

#8: Weak Analogy.

#9: Appeal to Pity.

#10: The Straw Man. Set up a fake argument and knock it down.

#11: *Non Sequitur.* The premises don't support the conclusion.

#12: Missing the Point. The premises point to a different conclusion.

#13: Red Herring. Finding other areas of agreement that have nothing to do with the conclusion that is being sought.

#14: Appeal to Questionable Authority.

#15: Hasty Generalization. Appealing to inductive logic with too few examples.

#16: Slippery Slope. A chain of reasoning ($A \Rightarrow B \Rightarrow C \Rightarrow D$) with weak links or too many links.

#17: Hidden Assumption. Question asked has an untrue assumption built into it.

Argument A: I know three people who liked eating hamburger with thousand island dressing. They all got brain cancer. Therefore, you shouldn't eat that combination.

Argument B: We belong to the same club. You and I have been out hunting together for years. We even like the same popcorn. Come with me to my kid's baseball game.

Argument C: I know it's true because I read it in the newspaper.

Argument D: If you do that, I'll go bankrupt.

Argument E: If you do that, you are nasty, mean, brutish, and short.

Argument F: If you do that, everyone will think you are the greatest.

Argument G: The real reason you don't want to marry me is because your wife doesn't like me.

Chapter Six
Predicate Logic

Fred put his wife's life insurance policy in his pocket. He figured he wouldn't be needing it for quite a while. After the stink of Coalback dissipated he could concentrate again on that field of flowers. The asters were pretty. Fred let A = "The asters are pretty" in the classical logic of proposition letters.

Then B = "The buttercups are pretty."

C = "The carnations are pretty."

D = "The daffodils are pretty."

He soon realized just using proposition letters from classical logic could get a little boring. There wasn't much you could do with A, B, C, and D and the connectives.

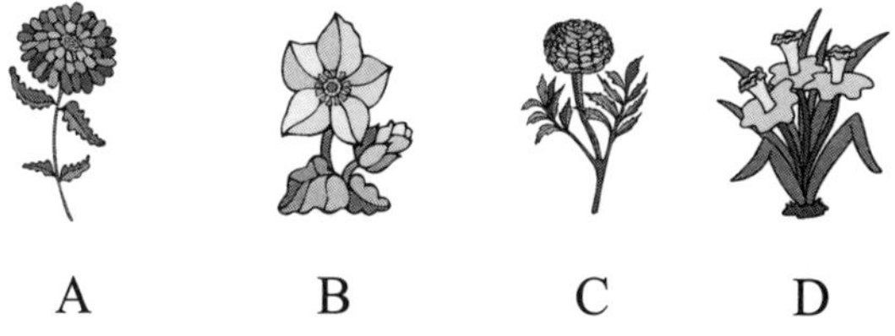

A B C D

Because classes were canceled on Monday, on Tuesday he would teach classical logic (= propositional logic = sentential logic).

On Wednesday he would expand the language L. He would introduce predicate logic (= first-order logic = quantifier logic).

Tuesday's classical logic language L contains just sentence letters, connectives, and parentheses.

Wednesday's predicate logic language L would also include four new items: ① predicate letters, ② constants, ③ variables, and ④ quantifiers.

Fred giggled. He imagined that on Tuesday he would write L on the board. On Wednesday, with the new expanded L, he would write **L**. Six-year-olds have such a juvenile sense of humor.

Instead of A = "The asters are pretty," B = "The buttercups are pretty," and so on, he would introduce **constants**. Let a = aster, b = buttercup, c = carnation, and d = daffodil.

He would introduce a **predicate** P that stood for "___ is pretty."

Pa would mean that asters are pretty.
Pb would mean that buttercups are pretty.
Pc would mean that carnations are pretty.
Pd would mean that daffodils are pretty.

Hold it! I, your reader, have an objection. In good old classical logic you had the sentence letters P, Q, R, ***etc., and now you've got predicate letters*** P, Q, R, ***etc. How in blazes am I supposed to tell which is which?***

Easy. Just look. Suppose I write (P ⇒ Qb) & (R ∨ Pc). Some of these are sentence letters and some are predicate letters. It's easy to spot the constants. They are all lower case letters. Let's get some practice in English . . .

Puzzle #75: In a single sentence explain why P is a sentence letter and why Pc is a predicate letter followed by a constant. (The math/logic part is easy. The hard part is writing a whole English sentence without making an error.)

In class on Wednesday Fred would write a big expression on the board . . .

((Q & Rc) ⇒ (Rd ∨ P)) ⇔ (Pa & Rac)

and have one of the students come to the board and circle all the sentence letters . . .

((Q & Rc) ⇒ (Rd ∨ P)) ⇔ (Pa & Rac)

Oops. You goofed, Mr. Author. That is an error. You have two constants after the predicate letter.

Ha! Ha! Hee! Hee! Ho! Ho! You'll never guess.

You mean . . . Rac ***is a predicate letter that has two constants that follow it.***

You guessed.

You gotta give me an example. I understand Pc ***would mean that carnations are pretty if*** P ***stood for "*** ___ is pretty,***" and*** c ***was the constant equal to carnation.***

First, let's understand that the predicate P (followed by one constant) is in pure logic (up in the clouds). When we make a model of L (and bring it down to earth), we might assign P to the meaning "___ is pretty." The constant c is in pure logic, and in a model of L we might assign c to carnation.

In another model of L we might assign P to "___ hates pizza" and c to your author, Stan. Then Pc would definitely be false.

Now to answer your question. Assign a to Alexander and b to Betty and c to clam-flavored iced cream .

Assign R to the predicate "____ loves ____."

Then Rab is true, but Rac is false.

Language Lesson:

P (= "___ is pretty") is called a 1-ary predicate.

R (= "___ loves ___") is called a 2-ary predicate.

Q (= "___ touched the ___ of ___ while they were in the ___ watching ___ boil in a pot on the stove") is a 5-ary predicate.

Qabcde is true if . . .

a = Alexander
b = arm
c = Betty
d = kitchen
e = spaghetti

Qabcde is probably not true if you change any of the constants. If you change e to boots it will be false.

If you change d to movie theater, Qabcde will be false.

Imagine how many stories you could write with a plot line Qabcde just by changing the constants! This is how many romance novels are written.

Puzzle #25: Sentence letters like P are really predicate letters! If P = "Portia is pretentious," we could say that P is a _____ predicate. Fill in the blank. Take a wild guess. Hint: It is not 98-ary.

(Two review sentences) In this predicate logic (= first-order logic = quantifier logic), we already know everything about four of the six items of L—predicate letters (0-ary, 1-ary, 2-ary, . . .), connectives, parentheses, and constants. All we need are variables and quantifiers.

(Preview of coming attractions) Variables and quantifiers have to be introduced at the same time. Once we have them, our language L will become much more powerful.

without variables and quantifiers

with variables and quantifiers

∀ is a quantifier that means "for all" or "for every."

Stop! You weren't telling the truth—and that's important in a logic book. You said, "Variables and quantifiers have to be introduced at the same time." Those were your very words. You talked about the quantifier ∀ and didn't mention variables. Shame on you!

I was just about to do this in my very next breath.

Not good enough. I want them at the same time.

Okay.

Quantifiers

∀ means "for all."
∃ means "there exists."

∀ is an upside down A, which comes from "for All."
∃ is a sideways E, which comes from "there Exists."

Variables

x, y, and z are variables.
If we need more, we might use u, v, or w.
If we need an infinite number of them, we could use x_1, x_2, x_3 and so on.

We'll use the first end of the alphabet for constants (a, b, c, . . .) and the back end for variables (x, y, z).

Puzzle #37: (easy) What if we need 24 constants? How do we avoid running into the land reserved for variables?

On Wednesday Fred would introduce variables and quantifiers with this model:

Let S be the 1-ary predicate "___ is a student at KITTENS University."

Let b be the constant Betty.

Let d be the constant Donald Duck.

Then he would point out that Sb is true and Sd is false.

Then he would write ∃x Sx. This would translate as, "There exists someone who is a student at KITTENS. That is true because Sb is true.

Then he would write ∀x Sx. This would translate as, "Everyone is a student at KITTENS" or "For all possible x, Sx is true." ∀x Sx is not true because Sd is not true.

Puzzle #4: Which of these are true?

A) ∃x ¬Sx

B) ∀x ¬Sx

C) ¬∀x Sx

D) ¬∃x Sx

Fred stood up and started walking. It was in those odd moments between sitting on a bench looking at a field of wild flowers and standing up that Fred would sometimes plan his classroom fun. He was trying to think of where he could get a teddy bear and a real bear to show his students on Wednesday the difference between classical logic (= sentential logic = propositional logic) and first-order logic (= quantifier logic = predicate logic).

Fred walked past the tennis courts. He imagined that someday he would be tall enough to play. When the tennis racquet is almost as tall as you are, it's difficult to swing it.

One of his students, Glen, was playing. She waved at Fred and yelled, "Hi!"

This is perfect for my predicate logic. I could let P be the 1-ary predicate "___ is playing tennis." The constants would then be people. If g = Glen, then Pg would be true. If f = me (Fred), then Pf would be false.

Let H be the 1-ary predicate "___ is holding a tennis racquet." Then Hg is true and Hf is false.

Puzzle #51: Which of these are true?

A) $\exists x\ Px$

B) $\forall x\ Px$

C) $\forall x(Px \Rightarrow Hx)$

D) $\forall x(Hx \Rightarrow Px)$

E) $\exists x(Hx \Rightarrow Px)$

This coming week in advanced algebra Fred was going to be teaching about Venn diagrams. To illustrate the fact that everyone who is playing must be holding a racquet, he would draw . . .

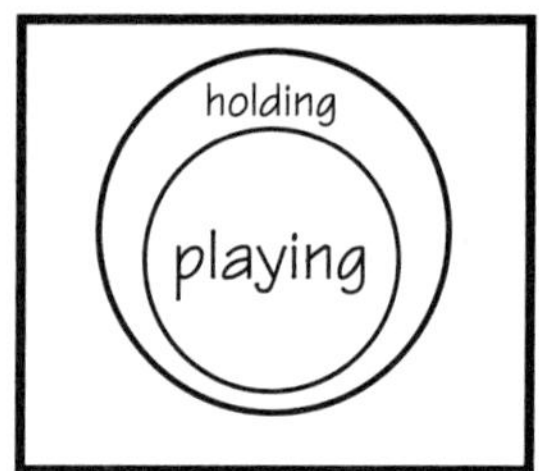

In logic he would write $\forall x(Px \Rightarrow Hx)$.

Puzzle #84: Draw a Venn diagram to illustrate $\neg\exists x(Ax \,\&\, Bx)$.

The most famous syllogism in logic is . . .

All men are mortal. Mortal = must die

Socrates is a man.

Socrates is mortal.

If we let M be the 1-ary predicate ____ is a man and
if we let s = Socrates and
if we let D be the 1-ary predicate ____ must die,
then this syllogism becomes $(\forall x(Mx \Rightarrow Dx) \,\&\, Ms) \Rightarrow Ds$.

In a Venn diagram, this syllogism would look like . . .

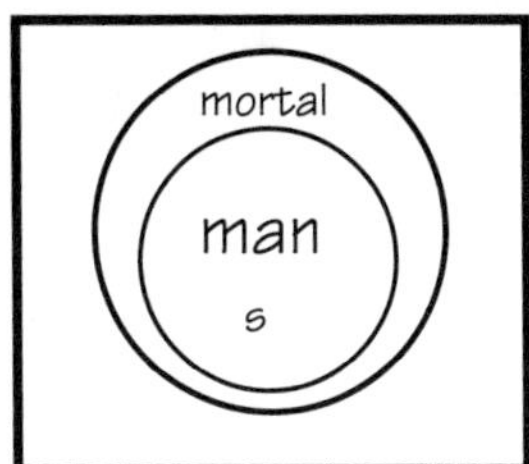

Since s is inside the inner circle, s is inside the outer circle.

$(\forall x(Mx \Rightarrow Dx) \ \&\ Ms) \Rightarrow Ds$ is true in predicate logic. It is true by the meaning of the symbols. This is the pure logic that lives in the sky. It will be true for *any* model (down on earth). It will be true for any assignment of M, D, and s.

The only teeny, tiny, little problem is that truth tables won't work for predicate logic. Fred's wonderful machine (from Chapter 3), which could be used to determine the truth of a propositional statement, is worthless for predicate logic.

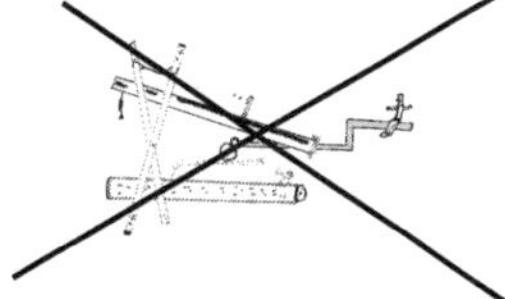

Here is why.

Why Fred's Machine Won't Work in Predicate Logic

In propositional logic there are no quantifiers. All we had to deal with are sentence letters. In pure logic we were given something like one of De Morgan's laws $\neg(P\ \&\ Q) \Leftrightarrow (\neg P \vee \neg Q)$. In order to show that it was always true (a tautology), we showed that it was true in every model of L. Namely, it was true for every assignment of truth values for P and for Q in applied logic. For two sentence letters there are four possible assignments.

Each structure of L will be a line in the truth table:

P	Q
T	**T**
T	**F**
F	**T**
F	**F**

For three sentence letters, there are eight lines in the truth table. (See page 39.)
If you had 4 sentence letters, there are 16 possible models. (See page 158.)

Number of sentence letters	Number of lines in the truth table	
1	2	**T F**
2	4	**TT TF FT FF**
3	8	etc.
4	16	

Puzzle #42: If you have n sentence letters, how many possible models would you need?

The important point is that for any expression in sentential (propositional) logic there are only a finite number of sentence letters and, therefore, only a finite number of different models that are possible.

When we allow the use of quantifiers things are different. Even the simplest expression like ∃x **Px** could have an infinite number of things to check in applied logic.

For example, suppose that **P** is the 1-ary predicate "____ is the smallest even integer greater than 2 that can't be expressed as the sum of two primes."

First, you check 4. $4 = 2 + 2$
Then check 6. $6 = 3 + 3$
Then 8. $8 = 3 + 5$
Then 10. $10 = 5 + 5$
and so on

In propositional logic, Fred's machine would only have to work for a finite amount of time before it announced, "This is a tautology" or "This isn't a tautology."

In predicate logic, any machine might grind on forever and never tell you Yes or No.

Fred was worried about his Wednesday class lecture. How was he going to establish the truth of something as simple as ∀x Ax ⇒ ∃x Ax, which is obviously true? "If Ax is true for all x, then certainly there exists an x for which Ax is true."

The happy answer is found in the next chapter, which is affectionately known as Chapter 7. But before we show how to prove that some expressions in predicate logic are true, we tidy up some loose ends here in predicate logic.

First tidy up: The symbol ⊨A for a sentence A in L still means that in every model for L, A is true. ⊨ applies to the down-to-earth applied part of logic. (page 43)

The symbol ⊢A for any sentence in L still means that A is true in the up-in-the-sky pure logic. (page 53)

The Tale of Two Worlds in which we talked about the castle of pure logic being up in the sky and the possible structures of L being on the ground was told starting on page 45.

Second tidy up: We can have more than one quantifier. Suppose we have the 2-ary predicate "___ is the mother of ___" where the constants are people.

Then ∀y∃x Mxy is true. It means that everyone has a mother.

Then ∃x∀y Mxy is silly. That means there is some mom that is the mother of everyone (including herself).

In contrast, in Chapter 7 we'll prove that ∀x∀y Axy ⇒ ∀y∀x Axy and we'll prove ∃x∃y Axy ⇒ ∃y∃x Axy.

Third tidy up: ∀x Bx means the same thing as ∀y By. The variables are just placeholders.

In algebra the distributive law was sometimes written as $a(b + c) = ab + ac$, but it could have been written as $x(y + z) = xy + xz$ or as $r(m + n) = rm + rn$.

The reflexive law could be written as a = b or as x = y or as ξ = ξ, using my favorite Greek letter ξ, which is pronounced xi. If you look up *xi* in any good English dictionary, it will tell you that ξ is the 14th letter in the Greek alphabet.

Fourth tidy up: Many other logic books talk about the **scope of a quantifier**. We may not need to use that concept a lot, but just for completeness sake, we'll show you what it means.

The scope of ∃x in the expression ∃x Px is $\exists x\ \underline{Px}$ (←what I have $\underline{\text{underlined}}$).

The scope of ∃x in ∃x(Px & Ry) ⇒ Px is $\exists x\underline{(Px \ \&\ Ry)} \Rightarrow Px$.

The scope of ∃x in ∃x∀y(Px & Ry) ∨ Qx is $\exists x \forall y\underline{(Px \ \&\ Ry)} \vee Qx$.

The scope of ∃x in ∃x∀y((Px & Ry) ∨ Qx) is $\exists x \forall y\underline{((Px \ \&\ Ry) \vee Qx)}$.

The scope of ∃x in ∀x(Px & ∃x(Pa & Rx)) is $\forall x(Px \ \&\ \exists x\underline{(Pa \ \&\ Rx)})$.

When there are no parentheses, it is straightforward. The scope is just the predicate that follows. ∃x Qabxc.

Many people can just "eyeball" what the scope is. They look at ∀x∀y∃z(Pab & ∃y((Pxy ⇒ Qax) ⇒ (Paz ∨ Pzb))) and can say that the scope of ∃y in that expression is . . .

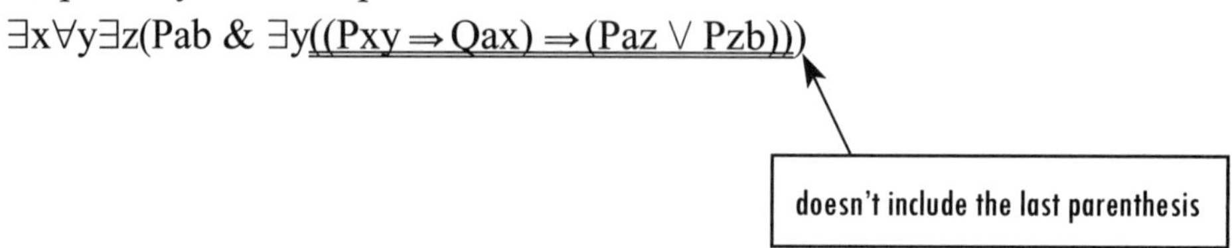

If Fred were going to build a machine to establish which expressions in predicate logic (= quantifier logic = first-order logic) are true, he would first need to have a mechanical way of determining the scope of each quantifier.

In computer talk we say that he would need an **algorithm**. An algorithm is a set of all the steps to accomplish something in a finite amount of time.

Algorithm to Find the Scope of ∃x in Any Expression

Case 1: If a sentence letter follows ∃x then the scope is that letter together with the constants or variables that follow it.

Case 2: For the first left parenthesis following ∃x start counting left and right parentheses. When the numbers match you have reached the end of the scope.

These are all the possible cases.

Here is the algorithm in action to determine the scope of ∀y in the expression ∀z∃y(Raz & Py) ⇒∀y∀z(((Py & Raz) ∨ (Py & Q)) ⇒ ∃x Pz)

123 1 4 23 4

I put the right parentheses numbers in gray.

Definition: y is **bound** iff it is in the scope of ∃y or in the scope of ∀y.

In the sentences we make in first-order logic, every variable will be bound. This ensures that every sentence will be either true or false. An expression like ∃x Rxy is nonsense. The variable y is not bound.

In some structure (= model) of L in which R is the 2-ary predicate "___ was born before ___ was born" and in which the constants are people, the nonsense expression ∃x Rxy would mean that someone was born before ___. I have no idea what that means.

Fifth tidy up: *English time!*

Watch me translate some basic English.

All cowboys ride horses. ∀x(Cx ⇒ Hx)

No cowboy is six thousand years old. ∀x(Cx ⇒ ¬Tx)

(Translation: If he's a cowboy, he's not 6000.")

Some cowboys are joggers. ∃x(Cx & Jx)

A common error is to write ∃x(Cx ⇒ Jx). To see why this doesn't represent "Some cowboys are joggers," let b = Betty. Since Betty isn't a cowboy, Cb is false. Then Cb ⇒ Jb is true by our definition of ⇒. (See the top half of page 32.) Then ∃x(Cx ⇒ Jx) is true because Betty isn't a jogger. That's awfully silly.

Some cowboys are not left handed. ∃x(Cx & ¬Lx)

Memory Aid: Associate *all* with ∀ and ⇒.
Associate *some* with ∃ and &.

Since this is English, expect things to get messy. If I want to express "All apples taste good," I would get ∀x(Ax ⇒ Tx).

If I want to say that all apples and bananas taste good, I really mean that if it's either an apple or a banana, then it tastes good.
∀x((Ax ∨ Bx) ⇒ Tx)

Note that ∀x((Ax & Bx) ⇒ Tx) is silly. Nothing can be both an apple and a banana at the same time.

Chapter Seven
Proving Things in Predicate Logic

Fred felt sad. He was thinking about his truth table machine—the machine that could decide whether a sentence in sentential logic was a tautology. That machine was great for Tuesday's lecture. Fred pictured that the janitor would come on Tuesday evening and throw his machine into the dumpster.

Fred was wrong.* His mood would brighten when he found out that his truth table machine would be an essential part of doing proofs in predicate logic, even though it couldn't prove directly the simplest first-order logic theorem $\forall x\ Ax \Rightarrow \exists x\ Ax$. (This translates as "If Ax is true for every x, then there is an x for which Ax is true.")

Here is the *Key Point* in the transition from classical logic (= propositional logic = sentential logic) to first-order logic (= predicate logic = quantifier logic).

In Tuesday's classical logic, a sentence S in language L was shown to be a tautology (translation: $\models$ S) by showing it was true in every model (= structure) of L. This was done by creating a truth table. All of this was done "on the ground" in applied logic.

☜ Applied logic and truth tables are down here.

In Wednesday's predicate logic, a sentence S in language L will be shown to be a tautology (translation: $\vdash$ S) by *proving* S is true in the castle of pure logic up in the sky. **(end of** *Key Point***)**

---

* How often are we sad because our beliefs are not rooted in reality?

When Fred taught geometry, all of his proofs were "up in the sky." When he proved that the base angles of every isosceles triangle were congruent, he didn't do that by drawing a thousand isosceles triangles and measuring their base angles.

Instead, he started with a "given" isosceles triangle ΔABC and then proved that $\angle B \cong \angle C$.

In geometry there were tons of postulates (such as "Two points determine a line) and tons of possible reasons you could offer for each statement you made (such as "given," "postulate," "previously proven theorem," "definition," "beginning of an indirect proof," "case 1," etc.)

Geometry students spend a year learning all the tricks of doing proofs.

Proofs in predicate logic will be a little different than geometry proofs, but they will be much less involved—unless, of course, you have been watching a lot of daytime television since the tenth grade and your brain has gone soft.

Do you remember lemmas from geometry? A **lemma** was a little baby theorem that you first proved before proving the main theorem. Then when you were proving the main theorem, you could tuck the results of the lemma inside the main theorem proof.

The real reason we first proved a lemma was that it made the proof of the main theorem shorter and more elegant.

If we didn't have lemmas, we could just prove the main theorem and incorporate inside of the proof the little subproof of the lemma.

Because proofs in predicate logic are often fairly short (under, say, five steps), we don't prove lemmas in advance of the main proof. Instead, we indent them over to the right side of the page.

There is a surprisingly short list of reasons that you can use in a predicate logic proof. Here's the list (and then I'll explain each one of them):

∀ elimination (known as ∀ elim)
∃ introduction (known as ∃ intro)
∃ elim
tautology (known as taut)
premise
from the lemma

∀ elim

If on some line of a proof you have ∀x Px, then on any later line you can write Pc where c is *any constant you like.*

Example: 8. ∀y Qy for some reason
. . .
27. Qb by ∀ elim from line 8

∃ intro

If on some line of a proof you have Ra, then on any later line you can write ∃x Rx.

Example: 5. Ac for some reason
. . .
73. ∃x Ax ∃ intro from line 5

∀ elim and ∃ intro have no restrictions.

∃ elim

If on some line of a proof you have ∃y Py, then on any later line you can write Pb where b is any constant *that has not been used.*

∃ elim must use a new constant.

Here is why.

I, your reader, was just about to ask that!

I beat you to the punch. ☺

You smiley-faced me!

I hope you weren't offended.

No. I kinda liked that. No other author has ever been so friendly to his readers.

Now, as I was saying, I want to explain why ∃ elim requires a new constant that hasn't been used previously in the proof.

Suppose we have on line 17 Sf where S is "___ is shorter than four feet tall" and f is the constant = Fred. Then Sf is true.

Suppose on line 23 we have ∃x Gx where G is "___ has worked in a grocery story for at least eight years."

On line 28, I can write Gb (where b has not appeared on any previous line), but I can't write Gf.

taut

Here is why Fred didn't have to be sad. Any expression that Fred's machine (or a truth table) could establish as a tautology is acceptable in a predicate logic proof.

Example:	4. R	for some reason
	5. S	for some reason
	6. R ∨ S	taut from lines 4, 5

The tautology used was (R & S) ⇒ (R ∨ S)

or, if you prefer,

$$\frac{R \quad S}{R \lor S}$$

or, if you prefer, R, S ⊨ R ∨ S.

Second example:	13. R	for some reason
	14. R ⇒ ∀x Qx	for some reason
	15. ∀x Qx	taut from lines 13, 14

The tautology used was (P & (P ⇒ Q)) ⇒ Q, which is *modus ponens*, the first tautology on the giant list on page 55.

Any sentence in the language of predicate logic is either true or false, just as any sentence in the language of propositional logic is either true or false. So all of the tautologies listed on page 55 will work when the sentence letters P is replaced by a sentence in predicate logic such as ∃y∀z(Axcy ⇒ Byddza).

Puzzle #95: Why is ∀x∃y(Px & Qy) ⇒ Px *not* a sentence in predicate logic?

In the olden days they made you memorize that giant list on page 55. But not today. And that giant list is just some of the "popular" tautologies. Instead, when you write a predicate logic proof, you can use any tautology you wish. The one who reads your proof is the one who has to resort to Fred's machine (or do a truth table) to ensure that you are playing fair.

The happy news is that about 95% of the time there are only a few tautologies that are commonly used.

premise

The first line of any lemma. Whenever you start a lemma, you can stick any sentence on the first line with the reason *premise*.

from the lemma

The results of a lemma.

If the first line of the lemma is P and the last line is Q, then back in the main proof, you can write P ⇒ Q and use the reason *from the lemma.*

Geep! This is a lotta stuff you just shoved at me. Six different possible reasons.

I know. But all the "shoving" is over. Once you see a couple of proofs, things will settle in. Once you have done some proofs, it will become routine. Before long, proving predicate logic theorems will be like making cookies or changing the oil in your car or sewing a button on a shirt. (← take your choice)

When we go from Bc to ∃x Bx, you will yawn and say ∃ intro. When we go from ∀x Ax to Ab, you will say ∀ elim.

It was back in geometry (when your mind was young and pliable) that you were faced with whiz-bang proofs—the proofs where you would exclaim, "How did they think of that!"

In geometry we proved that connecting the midpoints of *any* four-sided figure would give you a parallelogram. That proof involved a bit of creativity.

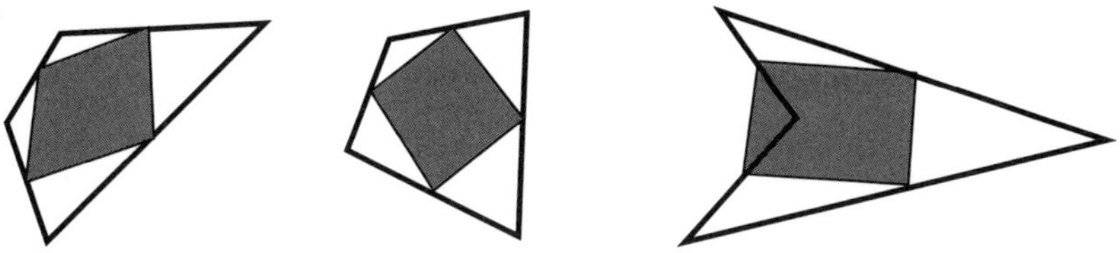

None of these logic proofs in this chapter will require more than 18 drops of sweat.

Here goes! Let's start with first-order logic theorem that was mentioned at the beginning of this chapter: ∀x Ax ⇒ ∃x Ax.

This could be written as ⊢ ∀x Ax ⇒ ∃x Ax.
Translation: ∀x Ax ⇒ ∃x Ax is true in pure logic.
Translation: ∀x Ax ⊢ ∃x Ax.

The proof . . .

a. ∀x Ax	premise
b. Ac	∀ elim
c. ∃x Ax	∃ intro

1. ∀x Ax ⇒ ∃x Ax   from the lemma

That's it. A one-line proof—assuming you don't count the lemma on the right hand side of the page.

Some notes . . .

♪#1: I'll number the lines of a proof with 1, 2, 3, and so on. I'll number the lines of the lemmas with a, b, c, and so on.

♪#2: On the first line of the lemma, I can write *any* expression I wish. The reason will always be *premise.*

♪#3: On line 1 of the proof, I took the first line of the lemma and the last line of the lemma and put a ⇒ between them.

Puzzle #33: Now it's your turn. Prove ⊢ ∀x Ax ⇒ Ac.

There were two tautologies on page 79 that I said that *we* would prove. The second one was the interchange of two ∃'s.

⊢ ∃x∃y Axy ⇒ ∃y∃x Axy.

The proof . . .

a. ∃x∃y Axy	premise
b. ∃y Aby	∃ elim
c. Abc	∃ elim
d. ∃x Axc	∃ intro
e. ∃y∃x Axy	∃ intro

1. ∃x∃y Axy ⇒ ∃y∃x Axy from the lemma

Some notes . . .

♪#1: For ∃ elim on line b of the lemma, I introduced the constant b. The rule for ∃ elim is that the constant introduced *may not have been used on any previous line.*

♪#2: The same is true when I introduced the constant c in line c.

That covers most of the first-order tautologies that I promised that we would prove.

Hold it! You forgot "the most famous syllogism in logic"—your very words—that you presented on page 76.

All men are mortal, and Socrates is a man. Therefore Socrates is mortal. You let M be "___ is a man," and let D be "___ must die," and let s = Socrates. In logic you wrote (∀x(Mx ⇒ Dx) & Ms) ⇒ Ds.

Okay. Now prove it.

With pleasure. I like doing these proofs.

The proof . . .

a. $\forall x(Mx \Rightarrow Dx)$ & Ms	premise
b. $\forall x(Mx \Rightarrow Dx)$	taut from line a
c. Ms $\Rightarrow$ Ds	$\forall$ elim
d. Ms	taut from line a
e. Ds	taut from lines c and d

1. $(\forall x(Mx \Rightarrow Dx)$ & Ms$) \Rightarrow$ Ds from the lemma

Puzzle #5: What tautology did I use to get line e?

Puzzle #39: (a little harder) What tautology did I use to get line d from line a?

small essay

Why We Don't Have a ∀ intro

This is review.....

We have a ∀ elim. If you have ∀x Qx on some line, then on any later line you can write Qa where a is any constant.

We have ∃ intro. If you have Qa on some line, then on any later line you can write ∃x Qx.

We have ∃ elim. If you have ∃x Qx on some line, then on any later line you can write Qc where c is any constant that has not appeared on any previous line.

In contrast, "∀ intro" presents all kinds of problems. If you have Kf on some line, you can't jump to the conclusion that ∀x Kx. In some structure (= model) K might be "___ has a doll named Kingie," and f might be the constant f = Fred. It's true that Kf, but it certainly isn't true that everyone has a doll named Kingie.

Even if you know Ka, Kb, Kc, Kd, Ke, and Kf, you still can't claim that ∀x Kx.

Even if you had an infinite number of true statements Ra_1, Ra_2, Ra_3, Ra_4, . . . it might not be true that ∀x Rx. For example, in a structure in which R is "___ is an even number," and in which $a_1 = 10$, $a_2 = 100$, $a_3 = 1000$, $a_4 = 10{,}000$, . . . it isn't true that ∀x Rx (since Rb is false when b = 7).

Some textbooks go through all kinds of contortions to create a ∀ intro rule. I won't describe the details. They are too horrible to imagine. We really don't need a ∀ intro rule.

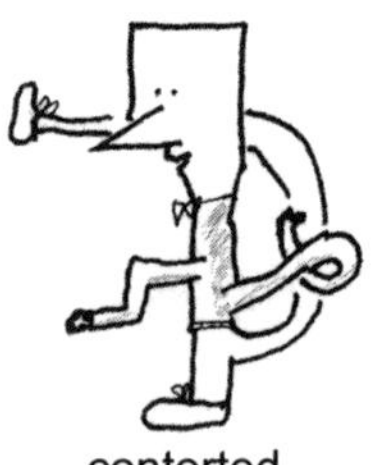
contorted

Instead, here is a nice alternative: Anywhere you see a ∀x in a proof, you can substitute ¬∃x¬ in its place.

Why is this true? Suppose you know that it is always true for any x that Px = Stan likes pizza at x o'clock. In symbols: ∀x Px. Then in the ordinary meaning of the words it is true that *it is not the case* that there is a time in which pizza doesn't taste good to Stan. That translates as: ¬∃x¬Px.

In a minute I'm going to prove both ⊢ ∀x Ax ⇒ ¬∃x¬Ax and ⊢ ¬∃x¬Ax ⇒ ∀x Ax. This is equivalent to . . .

⊢ ∀x Ax ⇔ ¬∃x¬Ax.

When you know both ⊢ P ⇒ Q and ⊢ Q ⇒ P, this is equivalent to . . .

⊢ P ⇔ Q.

Translation: P ⇒ Q, Q ⇒ P ⊢ P ⇔ Q.

end of small essay

⇔ is magic. When you know ⊢ P ⇔ Q, then P and Q have the same truth value: P is true exactly when Q is true. Then *anywhere* you see a P, you can substitute a Q.

Example:	7. ⊢ S ⇔ T	for some reason
	. . .	
	22. ∀x(S & (Px ⇒ Qxa))	for some reason
	. . .	
	39. ∀x(T & (Px ⇒ Qxa))	taut from lines 7, 22

So once I prove ⊢ ∀x Ax ⇔ ¬∃x¬Ax, I will be able to get around having to deal with the dreaded ∀ intro situation.

Okay. Prove it. I, your reader, understand you'll probably want to do it in two pieces: ⊢ ∀x Ax ⇒ ¬∃x¬Ax ***and its converse,*** ⊢ ¬∃x¬Ax ⇒ ∀x Ax. ***Fire away. I'm ready.***

I need one more second before I prove ⊢ ∀x Ax ⇒ ¬∃x ¬Ax. Please. I don't want to *rush things*.

There are basically two pictures of how logic proofs look. The proofs that we have done so far are **direct proofs**. They look like . . .

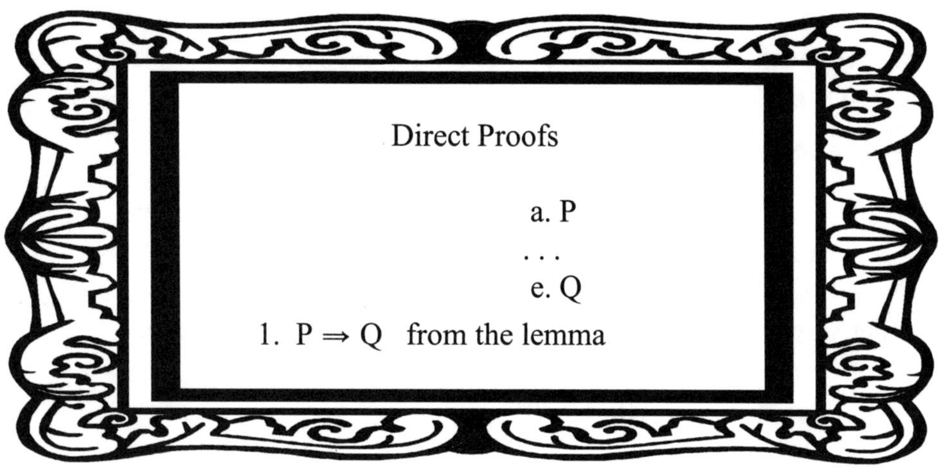

The other picture of logic proofs is the **indirect proof**. If you remember back a hundred years ago to your geometry class, we used to do indirect proofs. You assume the opposite of what you want to prove and then find a contradiction.

In logic here is the picture of an indirect proof of ⊢ P ⇒ Q . . .

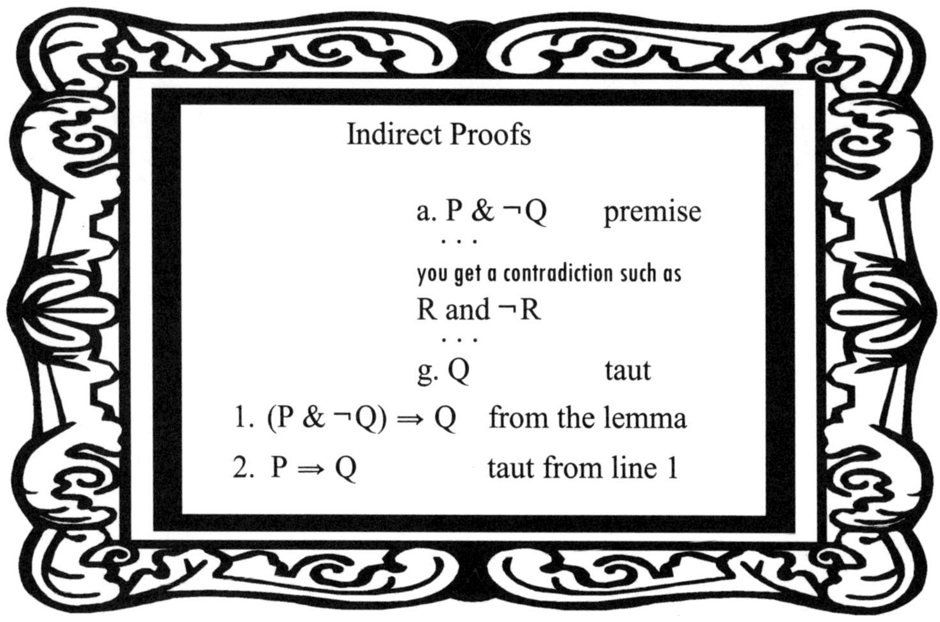

I have two tauts to explain in that picture of the indirect proofs. The first is from R and ¬R to Q.

$$\models (R \,\&\, \neg R) \Rightarrow Q.$$

If you start out with something false—such as R & ¬R—then *anything* is possible.

Puzzle #63: Prove ⊨ (R & ¬R) ⇒ Q. (Use a truth table.)

The second taut in the picture of indirect proofs takes us from (P & ¬Q) ⇒ Q to P ⇒ Q. In English this means that if P and ¬Q can together show that Q is true, then P alone can show that Q is true.

$$\models ((P \,\&\, \neg Q) \Rightarrow Q) \Rightarrow (P \Rightarrow Q).$$

Puzzle #98: Prove ⊨ ((P & ¬Q) ⇒ Q) ⇒ (P ⇒ Q).

These two tautologies were the last two in the giant list of tautologies on page 55. These two and *modus ponens* will be the workhorses for doing proofs in predicate logic.

Now, from two pages ago, you asked me to prove both ⊢ ∀x Ax ⇒ ¬∃x ¬Ax and ⊢ ¬∃x ¬Ax ⇒ ∀x Ax.

Combining those two together, ⊢ ∀x Ax ⇔ ¬∃x ¬Ax, will allow the "⇔ is magic" of substituting ¬∃x ¬Ax anywhere I encounter ∀x Ax. We will never need the ∀ intro rule.

I will need the indirect proof approach to prove each of these theorems.

Proof of ⊢ ∀x Ax ⇒ ¬∃x ¬Ax . . .

a. ∀x Ax & ¬¬∃x ¬Ax premise

When proving P ⇒ Q, the first step in an indirect proof is to start with P & ¬Q as the premise.

b. ∀x Ax taut from line a

c. ¬¬∃x ¬Ax taut from line a

The second and third lines of the lemma in an indirect proof are almost always P and ¬Q. The tautologies used are (P & ¬Q) ⇒ P and (P & ¬Q) ⇒ ¬Q.

These first three lines of any indirect proof are all automatic. If you are proving P ⇒ Q by the indirect approach, the first line is P & ¬Q. The second line is P. The third line is ¬Q.

You can be multitasking (reading the newspaper, watching television, exploring the Internet, *and* talking with your friend) and get these lines written. Better yet, give your baby brother a nickel and have him fill out these first three lines.

It's at this point that you have to stop multitasking and take the crayon out of your kid brother's hand and get to work. Your job is to find a contradiction. It usually looks like R on some line and ¬R on another line.

Once you have found the contradiction, you are back on automatic pilot. The rest of the proof is again mechanical.

Let's go back to the proof and get the contradiction.

d. ∃x ¬Ax   taut from line c

e. ¬Ac   ∃ elim from line d

Recall that when using ∃ elim, the constant that is introduced must be new to the proof. I haven't used the constant c earlier in this proof.

f. Ac   ∀ elim from line b

At this point you can turn off half your brain. You have a contradiction. You have three automatic lines yet to write. The first line is Q (if you are proving P ⇒ Q). This is the last line of the lemma.

g. ¬∃x ¬Ax   taut from lines e and f

Next is the first line of the proof. You take the first line of the lemma, put in a $\Rightarrow$, and then put the last line of the lemma.

1. $(\forall x\ Ax\ \&\ \neg\neg\exists x\ \neg Ax) \Rightarrow \neg\exists x\ \neg Ax$ from the lemma
2. $\forall x\ Ax \Rightarrow \neg\exists x\ \neg Ax$ taut from line 1.

The tautology used in line 2 is the one you proved in puzzle #98, two pages ago: $\vDash ((P\ \&\ \neg Q) \Rightarrow Q) \Rightarrow (P \Rightarrow Q)$.

❀ ❀ ❀

Now to prove the converse, $\vdash \neg\exists x\ \neg Ax \Rightarrow \forall x\ Ax$. This time I won't include all the interlarded commentary.

Proof . . .

a.	$\neg\exists x\ \neg Ax\ \&\ \neg\forall x\ Ax$	premise
b.	$\neg\exists x\ \neg Ax$	taut from line a
c.	$\neg\forall x\ Ax$	taut from line a
d.	$\neg\neg Ac$	$\exists$ elim from line b
e.	Ac	taut from line d
f.	$\neg Ac$	$\forall$ elim from line c
g.	$\forall x\ Ax$	taut from lines e and f

1. $(\neg\exists x\ \neg Ax\ \&\ \neg\forall x\ Ax) \Rightarrow \forall x\ Ax$ from the lemma
2. $\neg\exists x\ \neg Ax \Rightarrow \forall x\ Ax$ taut from line 1

❀ ❀ ❀

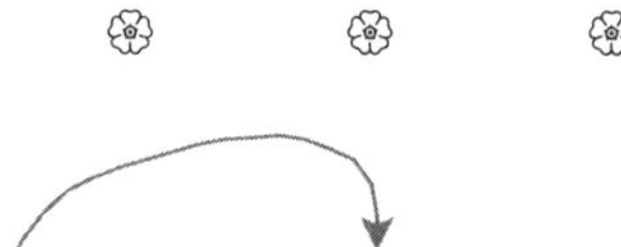

To officially prove $\vdash \forall x\ Ax \Leftrightarrow \neg\exists x\ \neg Ax$, all I have to do is to stack the two proofs I just did—one of them followed by the other. That will make a four-line proof. Line 2 will be $\forall x\ Ax \Rightarrow \neg\exists x\ \neg Ax$ and line 4 (after renumbering) will be $\neg\exists x\ \neg Ax \Rightarrow \forall x\ Ax$.

Then line 5 will be $\forall x\ Ax \Leftrightarrow \neg\exists x\ \neg Ax$ by taut from lines 2 and 4.

The tautology (fifth from the bottom on page 55) is $\vDash ((P \Rightarrow Q)\ \&\ (Q \Rightarrow P)) \Leftrightarrow (P \Leftrightarrow Q)$. Some people think of this as the definition of $\Leftrightarrow$.

We have now officially slain the $\forall$ intro dragon. With the magic of $\Leftrightarrow$, and the proven $\vdash \forall x\ Ax \Leftrightarrow \neg\exists x\ \neg Ax$, we can substitute $\neg\exists\neg$ anywhere we find $\forall$.

In the solution to puzzle #4 on page 149, I promised that we would prove . . .

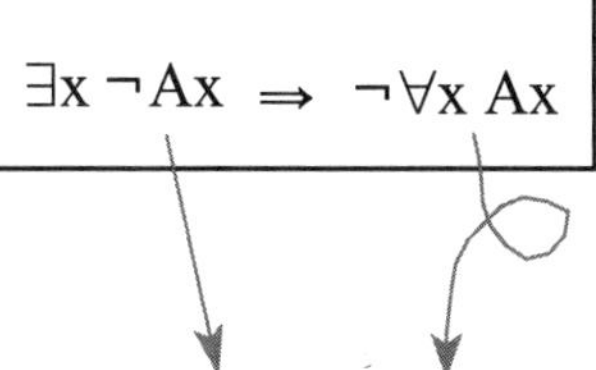

Here goes . . .

a. $\exists x\ \neg Ax\ \&\ \forall x\ Ax$ premise

You instantly recognize that I'm doing an indirect proof.

b. $\exists x\ \neg Ax$ taut from line a

c. $\forall x\ Ax$ taut from line a

e. $\neg Ac$ $\exists$ elim from line b

f. Ac $\forall$ elim from line c

g. $\neg\forall x\ Ax$ taut from lines e and f

1. ------

Puzzle #6: Finish up these last two lines of the proof. (Write them out on paper *before* you turn to my answers in the back of the book. The object of the game is to learn how to do indirect proofs—not to just get through the book!)

I just thought of something. I could prove $\exists x\ \neg Ax \Rightarrow \neg\forall x\ Ax$ without having to resort to an indirect proof.

A direct* proof of $\vdash \exists x \neg Ax \Rightarrow \neg\forall x\, Ax$ . . .

Two pages ago we finally finished proving $\vdash \forall x\, Ax \Leftrightarrow \neg\exists x \neg Ax$. Let's use that tautology.

1. $\forall x\, Ax \Leftrightarrow \neg\exists x \neg Ax$ — taut
2. $\forall x\, Ax \Rightarrow \neg\exists x \neg Ax$ — taut from line 1 $\vDash (P \Leftrightarrow Q) \Rightarrow (P \Rightarrow Q)$
3. $\neg\neg\exists x \neg Ax \Rightarrow \neg\forall x\, Ax$ — taut from line 2

   This is the contrapositive: $\vDash (P \Rightarrow Q) \Rightarrow (\neg Q \Rightarrow \neg P)$
4. $\exists x \neg Ax \Rightarrow \neg\forall x\, Ax$ — taut from line 3 $\vDash P \Leftrightarrow \neg\neg P$

There is only one more predicate logic theorem that I promised you that I would prove. On page 79 I promised that we would prove $\forall x\forall y\, Axy \Rightarrow \forall y\forall x\, Axy$, which shows that we could interchange two $\forall$'s.

The proof . . .

a. $\forall x\forall y\, Axy\ \&\ \neg\forall y\forall x\, Axy$	premise
b. $\forall x\forall y\, Axy$	taut from a
c. $\neg\forall y\forall x\, Axy$	taut from a
d. $\forall y\, Aby$	$\forall$ elim from b
e. Abc	$\forall$ elim from d
f. $\neg\forall x\, Axc$	$\forall$ elim from c
g. $\neg Abc$	$\forall$ elim from f
h. etc.	

Hold it! You are going to stop here? I, your reader, paid good money for this book. Please finish it up—and DON'T MAKE IT INTO A Puzzle FOR ME!

As you wish.

h. $\forall y\forall x\, Axy$	taut from lines e and g

1. $(\forall x\forall y\, Axy\ \&\ \neg\forall y\forall x\, Axy) \Rightarrow \forall y\forall x\, Axy$ — from the lemma
2. $\forall x\forall y\, Axy \Rightarrow \forall y\forall x\, Axy$ — taut from line 1

---

*This is so direct that we don't even need a lemma.

Suppose you had to sit in a chair and watch someone else paint a house. For the first hour it wouldn't be so bad, but after five or six hours, genuine boredom would set it. Every ten minutes in the chair would seem like an eternity.

In contrast, the time would go by much more quickly if you were allowed to do the painting yourself.

You, my reader, have had to do the harder job of watching me, your author, paint predicate logic proofs.

It's Your Turn to Paint.

Puzzle #52: $\vdash \forall x(Ax \Rightarrow Bx) \Rightarrow (\exists x\ Ax \Rightarrow \exists x\ Bx)$

Solving this puzzle may take anywhere from five minutes to an afternoon. This is not a horse race. (It took me 20 minutes.) Puzzles that begin with "$\vdash$" are a game—just a game. Happily, your life does not depend on winning this game within a certain time limit.

You know that you will start an indirect proof by writing the lemma that begins with P & ¬Q (if you are trying to prove $\vdash$ P $\Rightarrow$ Q).

You know that the second and third lines of that lemma will be b. P and c. ¬Q.

The first three lines are written before you break a sweat.

This puzzle—this game—is like many challenges in life. It will tend to either build your character or destroy you. Part of character is the quality of fortitude. Fortitude = the ability to stick to a task till it's done. We all start out in life with very little fortitude. With the slightest setback little kids will quit.

When we read the biographies of great men and women, it is often their ability to never never never give up that was an essential part of their greatness.

In other words, don't just read the puzzle, scratch your head for three minutes, and turn to my solution.

Intermission

Dealing with $\neg$

We have proved $\forall x\ Ax \Leftrightarrow \neg\exists x \neg Ax$.
Because it contains $\Leftrightarrow$, that means by what I call the "magic of $\Leftrightarrow$" that anywhere you spot a $\forall x$ you can substitute $\neg\exists x \neg$ in its place.

And anywhere you spot a $\neg\exists x \neg$ you could substitute $\forall x$ in its place.

On page 95 I have also proved the $\Rightarrow$ half of

$$\exists x \neg Ax \Leftrightarrow \neg\forall x\ Ax.$$

If I had done the proof of the converse, then anywhere you spot $\exists x \neg$ you could substitute $\neg\forall x$.

And then there are two additional tautologies:

$$\forall x \neg Ax \Leftrightarrow \neg\exists x\ Ax$$

and

$$\neg\forall x \neg Ax \Leftrightarrow \exists x\ Ax.$$

And the Law of Double Negation from page 55:

$$\models P \Leftrightarrow \neg\neg P.$$

Stop! Too many words! You are becoming a wordy-birdy. Give me the memory aid. Real simple please.

Okay. Replace $\forall$ with $\neg\exists\neg$ or vice versa.
Replace $\neg\forall$ with $\exists\neg$ or vice versa.
Replace $\exists$ with $\neg\forall\neg$ or vice versa.
Replace $\neg\exists$ with $\forall\neg$ or vice versa.

Replace $\neg\neg$ with nothing or vice versa.

In each case, you can use "taut" as the reason.

These replacements involving $\neg$ make this puzzle very straightforward. You can prove this puzzle as a direct proof.

Puzzle #90: $\vdash \neg\forall x\forall y\exists z\ Axyz \Rightarrow \exists x\exists y\forall z \neg Axyz$

My solution for this puzzle #45 will use an indirect proof and use some of the replacement rules for $\neg$ from the previous page.

Puzzle #45: $\vdash \exists x \forall y\ Axy \Rightarrow \forall y \exists x\ Axy$

The only thing left in proving predicate logic theorems are two minor variations.

First variation: We've done a lot of P $\vdash$ Q. (translation: $\vdash P \Rightarrow Q$) Now we'll deal with P, Q $\vdash$ R.

We want $(P \,\&\, Q) \vdash R$ to imply that $\vdash (P \,\&\, Q) \Rightarrow R$

Puzzle #76:

$$\frac{\begin{array}{c}\forall x(Ax \Rightarrow Bx)\\ \forall x(Bx \Rightarrow Cx)\end{array}}{\forall x(Ax \Rightarrow Cx)}$$

Given $\forall x(Ax \Rightarrow Bx), \forall x(Bx \Rightarrow Cx) \vdash \forall x(Ax \Rightarrow Cx)$

we want to show $\vdash (\forall x(Ax \Rightarrow Bx) \,\&\, \forall x(Bx \Rightarrow Cx)) \Rightarrow \forall x(Ax \Rightarrow Cx)$

Second variation: Let's look at $\vdash \exists x(Ax \vee Bx) \Leftrightarrow \exists x\ Ax \vee \exists x\ Bx$.

We need to prove both directions for a $\Leftrightarrow$ proof.

Going in the $\Rightarrow$ direction is a regular old proof.

Puzzle #8: $\vdash \exists x(Ax \vee Bx) \Rightarrow \exists x\ Ax \vee \exists x\ Bx$

One of De Morgan's laws will come in handy:

$\vDash \neg(P \vee Q) \Leftrightarrow (\neg P \,\&\, \neg Q)$ from the giant list on page 55

But the converse* $\vdash (\exists x\ Ax \vee \exists x\ Bx) \Rightarrow \exists x(Ax \vee Bx)$ is in the form $\vdash (P \vee Q) \Rightarrow R$.

Back in old geometry days, this was called **proving things by cases**. To prove $P \vee Q \vdash R$, you first show $P \Rightarrow R$. Then show $Q \Rightarrow R$. Then use the tautology $\vDash ((P \Rightarrow R) \,\&\, (Q \Rightarrow R)) \Rightarrow ((P \vee Q) \Rightarrow R)$. In English this might be modeled by If Peaches make you Rocky and if Quinces make you Rocky, then it must be true that either one will make you Rocky.

---

*The **converse** of $P \Rightarrow Q$ is $Q \Rightarrow P$.

Puzzle #71: ⊢ (∃x Ax ∨ ∃x Bx) ⇒ ∃x(Ax ∨ Bx)

You will need two lemmas. The first lemma will show that ∃x Ax ⇒ ∃x(Ax ∨ Bx).

That about wraps up proving tautologies in quantifier logic. At this point, I see little use in providing you with 30 more tautologies to prove.

➤ Detecting Invalid Arguments in First-Order Logic ◀

On the other hand, learning to detect invalid arguments in predicate logic might be more valuable to you in everyday life than proving valid arguments.

You proved puzzle #45: ⊢ ∃x∀y Axy ⇒ ∀y∃x Axy on the previous page. Back on page 79 in the "second tidy up" we showed that the converse to puzzle #45 isn't true.

In symbols: ⊬ ∀y∃x Axy ⇒ ∃x∀y Axy.

We did that by finding a structure (= model) in which the implication doesn't hold. We let Axy be the 2-ary predicate ___ is the mother of ____," and let the constants be people.

In that model ∀y∃x Axy ⇒ ∃x∀y Axy would mean that "Everyone has a mother" implies "Someone is the mother of everyone."

This is **T** ⇒ **F**, which isn't true.

Puzzle #60: Find a model in which ∃x Ax ⇒ ∀x Ax is not valid.

Puzzle #34: Find a model in which ((∀x(Px ⇒ Qx) & ¬Pc) ⇒ ¬Qc is not valid.

One technique for creating models to prove invalidity, is to make the models really simple. Restrict the **universe of discourse** (= the possible values of the variables) to one or two items.

For example, to show that

∃x Cx ⇒ ∀x(Ax ⇒ Bx), ∃x Ax ⊬ ∃x Cx ⇒∀x Bx

let the universe of discourse for x be {1, 2},
let A be "___ is equal to 2,"
let B be "___ is even," and
let C be "___ is less than 5."

Then both $\exists$x Cx $\Rightarrow$ $\forall$x(Ax $\Rightarrow$ Bx) and $\exists$x Ax are **T**.

If there exists a number less than five, then it must be true that 2 is even.

There is a number equal to 2.

But $\exists$x Cx $\Rightarrow$ $\forall$x Bx is **F**.

If there is a number less than 5, then it must be true that every number is even.

➤ end of Invalid Arguments in First-Order Logic ◄

Quick Recap . . .

Let S be any sentence in our logic language L. If S is provable in pure logic, we write ⊢ S. If S is true in every model of L, then we write ⊨ S.

When language L was only propositional logic (= sentential logic = classical logic), then we could determine whether ⊨ S by creating a truth table (or using the machine that Fred built). That truth table would have 2^n lines if S contained n different sentence letters. (Translation: The truth table never got infinitely large.)

When language L was expanded to predicate logic (= quantifier logic = first-order logic), then establishing ⊢ S was done by finding a proof for S. We had to find a proof for S, because there are an infinite number of models for even something as simple as $\forall$x Px $\Rightarrow$ $\exists$x Px.

Puzzle #96: Give a half dozen models for $\forall$x Px $\Rightarrow$ $\exists$x Px.

end of Quick recap

Fred knew that he was going to face two major questions near the end of his Wednesday lecture on predicate logic.* Both questions are hard to answer.

Fred had presented the rules for making predicate logic proofs:

∀ elim
∃ intro
∃ elim
taut
premise
from the lemma
the magic of ⇔ **

Some student will ask the First Question: Are these rules for doing proofs **sound**? (Translation: Are these rules goofy? Translation: Can I prove untrue things using these rules? Translation: Using these rules can I prove ⊢ P & ¬P? Translation: Are the rules **consistent**?)

Hey! You said this would be a hard question to answer. I, your reader, would have no trouble answering it.

You would have no trouble? What would you say?

I'd say, "Yes. The rules are sound." See how easy it is to answer.

If Fred were an authority figure—parent, judge, king, bishop, or KITTENS University administrator—he might get away with that. But he's only a undersized six-year-old with a pointy nose, a square head, and two little dots for eyes. Nobody is just going to take his word when he says that the rules he has presented are sound.

Here is the outline of the argument he will present.

Suppose we have a proof of ⊢ P & ¬P. That proof is a bunch of lines where the last line is P & ¬P. We know that last line is not true. So somewhere in that list of lines of the proof is the first line in which we have something that is false. Let's say that that line is line 17.

---

*Tuesday's lecture on propositional logic had an easy conclusion: He would roll in his truth-table machine, which could answer any question of tautology for Ps, Qs, Rs, and connectives ¬, &, ∨, ⇒ , and ⇔.

**If ⊢ P ⇔ Q, then anywhere you see a P you can substitute a Q.

That would mean that lines 1– 16 are all true.

How did we get line 17? It came from some previous true line(s). How did it come from those previous true line(s)? We used one of those seven rules of proof.

At this point Fred would go through each of the seven rules and show that if line 17 came from any of the previous lines, then line 17 had to also be true. For example, $\forall$ elim states that if we have $\forall x\ Sx$ on some previous line, then line 17 could be Sc where c is any constant. All of his students would nod yes.

The only more difficult parts would be the rules for *premise* and for *from the lemma*. I'm going to put a summary of his explanations in tiny type. Please feel free to skip over them.

When I said that we pick the first line of the proof that is false, I meant the first numbered line of the main proof. The line marked premise in a lemma is a lettered line of a lemma. That premise line could be false since you are allowed to put any statement on that line that you wish.

Suppose you put sentence Q on the first line of a lemma. At the end of the lemma you will have a sentence R. After the first step of the lemma you never use either *premise* or *from the lemma* as reasons. You only use those other reasons. And those other reasons logically take you from Q to R. If Q is true, then R will be true.

And that we acknowledge $(Q \Longrightarrow R)$ on the numbered line of the proof with *from the lemma.*

Hence, at no point in the numbered lines of a proof would we go from a true line to a false one. Q.E.D.

Some student will ask the Second Question: Are these rules that you have presented **complete**? (Translation: If a sentence Q—such as $\forall x(Ax \ \&\ Bx) \Rightarrow Bc$—is true in every model, will we always be able to prove Q? Translation: If $\vDash Q$, then will we be able to establish $\vdash Q$? Translation: If Q is semantically true, will it always be syntactically provable? Semantics and syntax are defined on pages 20 and 45.)

If the rules of doing proofs are complete, that means they are strong enough to prove any **logically valid** sentence (= true in every structure).

The Rules of Proof Are . . .
Sound if they are not goofy. →
← Complete if they can always do the job.

Soundness and completeness are converses of each other.
Soundness------------If provable, then true.
Completeness--------If true, then provable.

Fred was a little nervous about this second question of whether his deductive system for first-order logic could do the job. He knew that there would be only two minutes left in Wednesday's lecture hour.

He knew that the answer was yes. His deductive system was complete. The first person to prove that was Kurt Gödel. He did it in his 1929 doctoral thesis. Then he published it in 1930 with the snappy title "Die Vollständigkeit der Axiome des logischen Funktionenkalküls." Fred gave a copy of the 12-page proof to each student.

He said this is Gödel's famous Completeness Theorem.*

Student reactions . . .

Darlene: This thing has lousy spelling. I can't read a word of it.

Joe: There are little Fred eyeballs over the ä, ö, and ü.

Betty: A chance to practice my German! (She is working on her doctorate in math. Many universities require a reading knowledge of a foreign language—usually chosen among German, French, and Russian—as part of the math doctoral program.)

It was 5 p.m. on Saturday, and Fred had planned his Tuesday and Wednesday lectures—lectures that would finish all of propositional logic and predicate logic.

Some notes . . .
♪#1: Some high school and college teachers don't have their Wednesday lectures planned by the proceeding Saturday.
♪#2: Many teachers don't complete all of propositional and predicate logic in two day's teaching. In fact, they spend a whole semester doing what we have done in seven chapters.

---

*Gödel's more famous Incompleteness Theorem is coming later in this book.

♪#3: If you are a high school student, you may now claim that you have done all of an honors logic course. Translation: You end here. You are done.

♪#4: Starting with Chapter 8 we work at the college level. It's for those who are done with high school football games and senior proms and detention halls. Your prefrontal cortex—reasoning and judgement—is more developed.* You are better able to reason abstractly.

A function is a rule which associates to each element of the first set exactly one element of the second set. It is one-to-one if no two elements of the first set are mapped to the same element of the second set.

If a function is one-to-one, must the cardinality of the second set be larger than the cardinality of the first set?

♪#5: The dots over the ä, ö, and ü in German, which Joe called little Fred eyeballs, are called umlauts. They change the sound of those three letters.

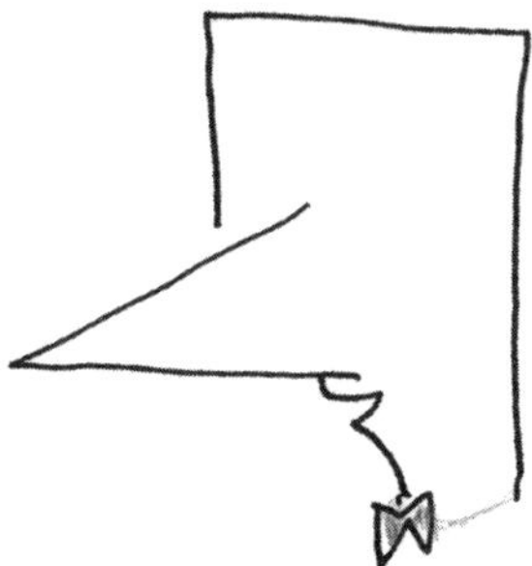

Fred without umlauts

*The development of the prefrontal cortex is completed by about age 23.

Chapter Eight
First Example of a Structure — Set Theory

Fred walked toward the campus library. He broke into song . . .

So much fun!
Logic is done!
Tautologies—both syntactic and semantic—
T Ts and proofs
With no goofs
It's all so terribly romantic.

In order to fit in the short line, Fred sang T Ts instead of truth tables.

The only word he could think of to rhyme to semantic is romantic. However, being 6, Fred didn't really understand romantic. He hadn't had many love affairs.

Last week when Fred had entered the library singing in his squeeky voice, the librarian showed him a copy of the library rules:

Library Rule #892.03c Six-year-olds are not permitted to sing in the library.

Fred finished his song before he got to the library. He didn't want to get in trouble again. He knew it was okay to smile in the library—as long as you did your smiling quietly.

Instead of heading into the library, Fred walked down the stone steps into the world famous Courtyard of Mathematics.

Over the years it had been filled with statues celebrating the many fields of math.

One statue was geometry. Another was algebra. Another was calculus. Set theory. Topology. Trig. Analysis. Arithmetic. Statistics. . . .

a theory

Each of these statues had the same shape. At the bottom were the axioms—the beginning assumptions. From the axioms the tree grew, consisting of definitions and proven statements (theorems). This whole thing—axioms, definitions, and theorems—is called a **theory**.

By one count, there are 128 subfields (theories) of mathematics.

Fred wandered through the Courtyard of Mathematics. Each of these statues was an old friend to him. He wondered whether he would get a chance to teach each of these at KITTENS in the next 80 or 90 years.

He had heard little children singing the alphabet song: A, B, C, D, E, F, G. . . . He wondered if he could invent a song with the 128 statues. He giggled: *This isn't the library. I can sing as loud as I want.*

The birds in the Courtyard of Mathematics starting inserting ear plugs as Fred began to sing.

Then a deafening silence. Fred realized that there was no statue for logic.* None.

That has to be a statue for logic! Fred thought. He checked all the math theory statues again and none of them was dedicated to logic. *This is terrible.* He sat down on the ground in the Courtyard of Mathematics and started to consider crying.

Then he looked down at the ground.

*In his head he expressed it as: The cardinality of the set of all logic statues in the Courtyard of Mathematics is equal to zero.

The cardinality of a set is the number of members in the set. card{❀, ☮} = 2

He looked closer. The whole Courtyard was paved with . . .

LOGIC.

Then Fred put two and two together* (an idiom) and realized that every statue rested on logic. Translation: Every subfield of math has logic as its foundation. Translation: Every subfield of math is a model in the language L of logic.

This is wonderful Fred thought. Everyone knows that first-order logic is both sound and complete (= if you can prove it, then it's true AND if it's true, then you can prove it).

What better basis for all of math.

Fred dreamed of someday writing a book on logic.

In Chapter 8 of my book, I will take some subfield of math—say, set theory—and show how it is a structure of L. That means that I will need to show what each predicate of L means and what each constant of L means.

---

*4

Set theory needs two predicates. One of them represents "is a member of." Set theorists will write b ∈ D to indicate that b is a member of set D. Our 2-ary predicate will be M.

b ∈ D is written as Mbd. In keeping with the first seven chapters of this book, the predicates will be capital letters and the constants will be lowercase letters.

The only other predicate will be equality (=). Equality (also known as identity) will be used in almost all the math models. Set theorists will write A = C to indicate that two sets are equal. Our 2-ary predicate will be E.

A = C will be written as Eac.

All the other predicates of L we'll assign to the garbage can.

Now all the things we know about logic will apply to M, to E, and to a, b, c, d. . . .

For example you could prove that ∀x∀y Mxy ⇒ ∃x∃y Mxy.

Hey! But where's the set theory? This is just logic stuff.

In order to get set *theory* we have to add the axioms of set theory. (See the statue from two pages ago.) We have to add the specific properties that M and E have.

The **axioms for equality** will be the same for all the models of L. We know these four things about equality:

Reflexive axiom: ∀x Exx

often written as a = a.

Symmetric axiom: ∀x∀y Exy ⇒ ∀x∀y Eyx

often written as a = b implies b = a.

Transitive axiom: ∀x∀y∀z((Exy & Eyz) ⇒ Exz)

often written as a = b and b = c imply a = c

Substitution axiom: If Ebc, then in any sentence containing b you can substitute c in place of b and not change the truth of the sentence.

For example if b = c and we know Mbe, then we can write Mce.

Perhaps, the most popular axioms for set theory are the Zermelo-Fraenkel axioms. Ernst Zermelo started creating them in 1908. Then Abraham Fraenkel, Dimitry Mirimanoff, and Thoralf Skolem made some suggestions/improvements. Only Ernie and Abe get credit in the naming of the Zermelo-Fraenkel axioms. Dimi and Thor got left out. I'm not sure why.

These eight axioms give us additional properties about the two binary (2-ary) predicates E and M. With them we can do a great deal of set theory.

Here are some of the axioms . . .

ZF#1—known as Set Equality

In English: Two sets are equal if and only if they have the same members.

In the language of set theory: $\forall x(x \in A \text{ iff } x \in B) \text{ iff } A = B$.

In the language of L: $\forall u \forall v(\forall x(Mxu \Leftrightarrow Mxv) \Leftrightarrow u = v)$.

ZF#3—known as the Axiom of Pairing

In English: If C and D are sets, then there is a set that contains both C and D as members.

In the language of set theory: $\forall C \forall D \exists A(C \in A \ \& \ D \in A)$.

Puzzle #11: Write ZF#3 in the language of L.

ZF#4—known as the Axiom of Union

In English: For any set A, there exists a set G that consists of all members of members of A.

Translation: If you have, for example, a set A that equals {{✿,☎}, {7, @}, {❍,✎,✇}}, then there is a set {✿,☎, 7, @, ❍,✎,✇}.

In *Life of Fred: Five Days* Fred spends a week (five lectures) presenting all eight Zermelo-Fraenkel axioms. Since it will take Fred only two days (Tuesday and Wednesday) for him to present all of propositional and predicate logic, you can imagine how much set theory Fred covered in those five lectures.*

---

*And those five lectures are the first five lectures of *a whole semester* of set theory.

At the end of his fifth lecture in set theory, Fred presents the Schröder-Bernstein theorem. This theorem uses the concept of **functions**.

It states that for any two sets, A and B, if there are functions f and g such that f:A → B and g:B → A that are both 1-1, then there must exist a 1-1 and onto function h:A → B. (It is proved on pages 193–198 in *Life of Fred: Five Days*—a big proof.)

The concept of function is used in almost every math subfield. The definition is simple enough: Given two sets, A and B, a function f is a rule that assigns to each element of A exactly one element of B.

If a is assigned to b, then we write f(a) = b.

The challenge is to define when a binary (2-ary) predicate is a function. Here's one definition that will do the job:

Definition: Binary predicate F is a function iff

$$\forall x \exists y(Fxy) \ \& \ \forall x \forall y \forall z((Fxy \ \& \ Fxz) \Rightarrow y = z).$$

Notation: If F is a function and if Fxy, then we write f(x) = y.

This says that f(x) has at least one assignment.

This says that f(x) has at most one assignment.

We need the fancy f, to indicate that this represents a predicate that is a function.

As we build set theory on top of logic and the axioms of set theory, all we need do is create theorems and definitions. For example . . .

Definition of the **union of two sets**: Set C is the union of set A and set B iff every element of C is in either A or B or both.
Notation: A ∪ B = C.
Translation into logic: $a \cup b = c \ \Leftrightarrow \ \forall x((Mxa \vee Mxb) \Leftrightarrow Mxc)$.

Nice Things You Can Say about Some Axiom Systems

Consistent. That means you can't prove contradictions such as P & ¬P.

(consistent = sound)

Complete. That means you can prove every true statement. Gödel's Completeness Theorem published in "Die Vollständigkeit der Axiome des logischen Funktionenkalkül" proved that logic is complete. Several chapters from now we will talk about whether the axioms of set theory or of arithmetic are complete.

Meaningful. It's really comforting if the axioms represent something and are not just some random assortment of statements. Classifying an axiom system as meaningful is subjective. What looks meaningful to one person may seem like gobbledegook to another.

Independent. That means that you can't prove any of the axioms from the other axioms. This is nice but not necessary.

Recursive. That means that you can tell whether any given sentence is an application of one of the axioms. This is not merely nice; it is necessary. There has to be an algorithm (some fixed procedure that is doable in a finite number of steps) to determine whether, say, the sentence on line 39 is an application of axiom 14.

It turns out that less than 1% of all the axiom systems for all the 128 statues of subfields of mathematics have all five nice attributes: consistent, complete, meaningful, independent, and recursive.

I'd say why, but I can't because this is the end of Chapter 8.

Chapter Nine
Last Example of a Structure—Arithmetic

Fred loved standing in the Courtyard of Mathematics. He walked over to the set theory model and "beeped" the nose of the lion at the base of the statue. It was a friendly beep and the lion didn't seem to mind at all.

He decided to go around and do that to every one of the statues. That would show his solidarity with all 128 subfields of math.

When he got to arithmetic, he touched the lion's nose and counted 1, 2, 3, 4, 5, 6. . . . Arithmetic is special because that's the first statue that kids meet when they explore the Courtyard of Mathematics. They spend years and years there . . .

learning to count

learning the addition tables

learning the multiplication tables

learning how to divide $5\overline{)280}$ (56)

learning how to add fractions

learning how to subtract fractions

learning how to multiply and divide fractions

learning what decimals are

learning how to add, subtract, multiply, and divide decimals

learning what percents are

learning how to change decimals into percents and vice versa

learning how to change decimals into fractions and vice versa

learning how to compute a 15% discount

learning how to find 6% of 30

learning how to find what percent 8 is of 24.

Kids spent so much time playing at the arithmetic statue that many think that arithmetic IS math. They can't imagine why anyone would want to have a career in yucky math. It would be like spending 40 years balancing your checkbook.

If music were presented in this way, kids would start in kindergarten having to

learn the difference between a semibreve and a treble clef

learn what an augmented chord is

learn . . .

Instead, kids like music because in kindergarten they get to beat on sticks or hit bells or sing songs about barnyard animals.

People pay money to go to music concerts or see art exhibitions or watch movies.

Music can be beautiful.

Art can be beautiful.

Movies can be beautiful.

Math can be beautiful.

The big difference is the amount of training/education needed to appreciate each of these arts.

The axioms for arithmetic are easier than the ones for set theory. A little over a hundred years ago (1899) Giuseppe Peano (pronounced piano) created the five axioms. They are called the Peano postulates. (*Postulate* means the same thing as *axiom*, but Peano postulates is much more alliterative than Peano axioms.)

Giuseppe was Italian but wrote his postulates in Latin. Today more people study computer languages than study Latin. Here they are in English . . . *After you quickly read them, I'll do some explaining.*

Peano postulate #1: 1 is a natural number.
Peano postulate #2: For every natural number n, **S**n is also a natural number.
Peano postulate #3: 1 is not the successor of any natural number.
Peano postulate #4: No two different numbers have the same successor.
Peano postulate #5: Suppose you have a set K and you know two things:

(i) 1 is in K and

(ii) for every natural number n, if n is in K, then **S**n is also in K, then K is the set of all natural numbers.

Peano postulate #1: 1 is a natural number.
Translation: The numeral 1 is a constant in the language of logic. All of the constants of L in the arithmetic model are natural numbers. We pick out one of them—1—as a special one.

Peano postulate #2: For every natural number n, **S**n is also a natural number. **S** is called the **successor** function.
Translation: We have a function *f*, which in arithmetic is called **S**. All functions are binary predicates by definition of "function." This postulate states that *f*(x) is also a number. When you see **S**n, think "n + 1."

Peano postulate #3: 1 is not the successor of any natural number.
Translation: $\neg\exists x(\mathbf{S}x = 1)$

Peano postulate #4: No two different numbers have the same successor.

Puzzle #31: Translate this ↑ Peano postulate #4.

Peano postulate #5 is **math induction** which is usually studied in second year high school algebra.

In five days of lecturing (in *Life of Fred: Five Days*) Fred starts with the Peano postulates and starts to construct the theory of arithmetic (the arithmetic statue). Building the theory means creating definitions and theorems based on the axioms.

He also defines a binary (2-ary) function "A" that behaves like addition. He also defines 2, 3, 4, 5 . . . as **S**1, **S S**1, **S S S**1, **S S S S**1. . . .

He proves Theorem 29: A2,2 = 4, which elementary school kids know as 2 + 2 = 4. He *proves* it—not just gives examples of why it's true.

Intermission

When I was in elementary school the teacher took two marbles and put them in her pocket. Then put two more marbles in her pocket. Then she pulled four marbles out of her pocket.

One of the kids put two lumps of clay in his pocket and then put two more lumps of clay in his pocket and "proved" that 2 + 2 = 1.

Fred proved the commutative law of addition: n + m = m + n, which, in logic, might be written as ∀x∀y(Axy = Ayx).

He defined the binary function M for multiplication. He proved the distributive law: a(b + c) = ab + ac. He defined < and exponents.

He defined integers. He defined addition and multiplication of integers and proved that the product of two negative integers is always positive.

He defined the rational numbers and showed that +, –, ×, and ÷ all work as they should.

Getting the real numbers from the rationals involved some fancy dance steps involving partitioning the rational numbers in a creative way so that there were over a million real numbers for every rational number.*

What is amazing is that all of these aspects of arithmetic could be derived from those five little Peano postulates.

You can see why the theory of arithmetic holds a special place in Fred's heart.

In Language ℒ of First-Order Logic

Connectives: ¬, &, ∨, ⇒, ⇔

Sentence letters: A, B, C, P_1, P_2, P_3 . . . (0-ary predicates)

Constants: a, b, c, d, a_1, a_2, a_3 . . .

Variables: x, y, z, x_1, x_2, x_3 . . .

Predicates Ac (1-ary), Bdc (2-ary), Cabcde (5-ary) . . .

Quantifiers: ∀, ∃

In the Arithmetic Model of ℒ

Assign a to 1.

Assign binary predicate S to ***S*** where ***S*** is a function. (See page 111 for the definition of function.)

The axioms of this model are the Peano postulates.

Think of the constants and variables as natural numbers {1, 2, 3, . . .}.

*Actually, for every rational number there are over 1000000000000000000000000 00000000000000000000000000000000000000000000000000000000000000000 0000000000000000000000000000000000000000000000000000 real numbers.

In the land of logic things are delightful. You can prove every true statement (complete), and you can't prove false statements (consistent).

In the arithmetic model of L, we can say . . .

$2 + 2 = 4$ $\frac{5}{8} \div \frac{3}{4} = \frac{5}{6}$ $2^3 < 3^2$ 6 divides evenly into 72

53 is a prime number   8 is 33⅓% of 24

. . . and we know in our experience that we have consistency.

BUT . . .

In the 1800s it was the Hope that the arithmetic model would be complete—that you could, in theory, prove every true statement.

In 1905 it was the EXPECTATION that the arithmetic model would be complete.

In 1930 everyone knew in their Heart that the arithmetic model is complete. All we had to do is be smart enough to establish that fact.

In 1931 . . .

Fred will never ever build a machine that will be able to decide the truth or falsity of every math statement . . .

. . . even in worlds as simple as arithmetic.

All of truth will never be captured/locked up/determined by any computer, any proof scheme, any algorithm that we will ever create. Not even in the little world of arithmetic. This was *proved* in 1931.

The rest of this book will prove this surprising fact.

Chapter Ten
Numbering Every Sentence

When Fred first learned about Gödel's Incompleteness Theorem (1931), which proved that mathematics that is at least as complicated as arithmetic will never be tamed—will always be wild—he didn't know whether to be happy or sad. What this theorem means is that robots will never be able to conquer all of math.

We will never be able to find a procedure that will work with every arithmetic statement to determine whether it's true.

We can prove a lot of things, such as 2 + 2 = 4 and that there are an infinite number of prime numbers.* But there will always be unprovable assertions in math.

This is nuts! I, your reader, want my money back. I understand that certain statements, such as Goldbach's conjectures that you mentioned in the footnote on page 19, might be hard to prove. But you are telling me that there are things that we will NEVER be able to prove!

Don't you realize that computers are getting smarter and smarter every month? How can you predict a hundred years into the future? I refuse to accept this silliness of unprovability just because you say it.

Do you mean that proving that certain true statements are forever unprovable sounds ridiculous?

You bet! That can't be done.

That's what billions** of mathematicians said in 1931 when they heard the rumor that Gödel had proved certain statements are never going to be proved.

And what did those hundreds of mathematicians say after they read the proof of Gödel's Incompleteness Theorem?

---

*You knew that, didn't you? There's a really old proof that there are an unlimited number of primes. Start by assuming there is only a finite number of primes. Call them p_1, p_2, p_3, . . . p_n. Their product $p_1p_2p_3 \ldots p_n$ is divisible by every prime.

Then $(p_1p_2p_3 \ldots p_n) + 1$ is *not divisible* by any prime and hence, must be a prime. This is a new prime that is not on the list of all primes. Contradiction.

**Speaking hyperbolically.

First, there was a stunned silence that lasted about a half an hour (Rev. 8:1). Then they realized that ancient truth that every generation must relearn . . .

Reality is quite often very unreal.

❂ Some early Greeks (Pythagorus) *knew* that every number is rational.*
❂ Everyone used to *know* that the sun goes around the earth. Some people still think it does.
❂ Everyone, until they get to advanced algebra, *knows* that you can't take the square root of a negative number.
❂ Everyone *knows* that a teenage girl with no education is not the one to lead conquering armies. (Joan of Arc)
❂ Everyone *knows* that there is no limit to what computers can achieve in mathematics.

I remain an unbeliever. Unless you can show me an unprovable true piece of arithmetic, unless I see it with my own eyes and wrap my brain around it, I will not believe.

Okay. Stay tuned. But first I have to do a little Gödel numbering.

We are going to assign a number to each character in the logic language L. In technical words, we'll find a function from the characters of L to the natural numbers ℕ (= {1, 2, 3, 4, . . .}.)

First, we'll do the connectives. Assign ¬ to 1. Assign & to 2. Assign ∨ to 3. Assign ⇒ to 4. Assign ⇔ to 5.

Assign right parenthesis to 6 and left parenthesis to 7.

¬	&	∨	⇒	⇔	(	)	∀	∃	**S**	1	P_1	c_1	x_1	P_2	c_2	x_2	. . .
1	2	3	4	5	6	7	8	9	10	11	12	13	14	15	16	17	. . .

That assignment covers all the connectives, the parentheses, the quantifiers, the successor function, the constant "1," and allows us an infinite number of predicates, constants, and variables.

---

*Rational numbers are numbers that can be expressed as an integer divided by an integer. 3/5, 89.3, 5%, –8/9, 0, etc. π and $\sqrt{7}$ are not rational.

Let's take that infinite list of assignments and trim it down a bit. In Chapter 3 ("Making L Simpler") we found we could replace all five connectives with $\downarrow$.

We could throw out A, B, C, . . ., X, Y, Z and just leave P_1, P_2, P_3, P_4. . . .

We could throw out the constants a, b, c, d, e and just leave c_1, c_2, c_3. . . .

We could throw out the variables x, y, z and just leave x_1, x_2, x_3. . . .

Then we replaced P_1 with $P_\blacklozenge$. Replaced P_2 with $P_{\blacklozenge\blacklozenge}$. Replaced P_3 with $P_{\blacklozenge\blacklozenge\blacklozenge}$, etc.

The c_1, c_2, c_3 . . . could be replaced with $c_\blacklozenge$, $c_{\blacklozenge\blacklozenge}$, $c_{\blacklozenge\blacklozenge\blacklozenge}$. . . .

The x_1, x_2, x_3 . . . could be replaced with $x_\blacklozenge$, $x_{\blacklozenge\blacklozenge}$, $x_{\blacklozenge\blacklozenge\blacklozenge}$. . . .

Tell me the truth. You didn't think we would ever use that "Making L Simpler" material again, did you?

Now we have a finite list of symbols in L.

New revised assignment list . . .

$\downarrow$	(	)	$\forall$	$\exists$	$\mathcal{S}$	1	P	c	x	$\blacklozenge$	
1	2	3	4	5	6	7	8	9	10	11	Nice & short!

At this point we can translate any sentence $\mathcal{S}$ of L into a sequence of letters. $\forall x \neg P_1 x$ would become

$\forall x \neg P_\blacklozenge x$ which would become

$\forall x (P_\blacklozenge x \downarrow P_\blacklozenge x)$ since $\neg P \Leftrightarrow (P \downarrow P)$ from page 42 of Chapter 3

and finally

4, 10, 2, 8, 11, 10, 1, 8, 11, 10, 3.

Ha! I caught you again. I, your reader, am not some dumb bunny. The title of this chapter is "Numbering every sentence." You, Mr. author, numbered every symbol of* L*. You started with a sentence* $\forall x \neg P_1 x$ *and got

a sequence of numbers, 4, 10, 2, 8, 11, 10, 1, 8, 11, 10, 3. ***You promised a single number for each sentence—not a sequence of numbers.***

I wasn't done. A little patience please. All I have to do now is create an assignment from any finite sequence of numbers, such as 4, 10, 2, 8, 11, 10, 1, 8, 11, 10, 3, to a single number.

I want that assignment (that function) to take different sequences to different numbers. I.e. that function will be one-to-one.

Why?

If that function is 1-1, then that function will have an inverse. Translation: I will be able to go from that single number back to the sequence that created it.

This is a neat trick that Kurt* did: finding a function from all finite sequences to single numbers that has an inverse.

Okay. I'm ready.

I just wanted you to appreciate what he did.

First, write out a list of the prime numbers: 2, 3, 5, 7, 11, 13, 17, 19, 23, 29, 31, 37, 41. . . .

Now take that sequence 4, 10, 2, 8, 11, 10, 1, 8, 11, 10, 3 and use them as exponents on those prime numbers:

$2^4 3^{10} 5^2 7^8 11^{11} 13^{10} 17^1 19^8 23^{11} 29^{10} 31^3$.

That product is a single number. And it is slightly huge. The point is that it can be multiplied out if necessary. All I needed to show is that each sentence $\mathcal{S}$ of L can be turned into a single number.

No. You also claimed that you could take that single number and turn it back into the sentence $\mathcal{S}$.

You are right. I need to show that that **Gödel number** can always be turned back into the sentence $\mathcal{S}$. This has to be a `mechanical` procedure (an algorithm).

I start with some Gödel number such as 907979239926396903213 5135135132500558443151513622600000.

I keep dividing it by the first prime, 2. That shrinks the number down. Suppose 2 divides into that big number eight times (until the number turns odd). I now have 2^8.

---

*Kurt Gödel, as he is more formally known.

Then I shrink the number down even further by dividing by the next prime, 3, as many times as I can. Suppose I can divide by 3 seventeen times. Then 3^{17}.

I continue until that 907979239926396903213513513513250055844315151362260000 is shrunk down to 1.

Now every sentence $\mathcal{S}$ of L has a unique Gödel number assigned to it. Since we are working in the arithmetic model of L, which is based on the Peano postulates, we can (in the model) do all of the multiplication, exponentiation, etc. necessary to find the Gödel number of any $\mathcal{S}$, and we can go from that Gödel number back to $\mathcal{S}$.

This is the first step in proving that the arithmetic theory is not complete—that you can't prove every true statement.

The second step is to turn every *proof* in the arithmetic model into a single number.

What's a proof? In *any* theory it is a list of statements in which (1) every statement has a reason that justifies it, and (2) the last statement is the thing you want to prove. Recall the proofs you did in high school geometry.

Statement	*Reason*
1. ΔABC with AC = BC	1. Given

The thing we never tell our geometry students is that the Reason column is unnecessary! To give a proof, all you need to do is give a list of statements.

If the statement on line 6 of a proof is $\angle 3 \cong \angle 5$, it is always possible to `mechanically` determine what the reason is. There is only a finite number of possible reasons, and you can check them one-by-one to see which one works.

You can check if it is one of the axioms of the theory.

You can check if it is from logic.

You can check if it is a definition.

You can check if it is a previously proven theorem.

He lists his reasons

The only reason that we list reasons in a proof is because we are nice, decent, caring mathematicians who love our readers.

So the second step in proving that the arithmetic model is incomplete is assigning a Gödel number to any finite *list* of statements. There are two ways to do that: the **quick and dirty** and the *Elegant*.

Quick and dirty: In the assignment list

↓	(	)	∀	∃	**S**	1	P	c	x	♦
1	2	3	4	5	6	7	8	9	10	11

add a 12th entry to separate the statements from each other. That new symbol could be something fancy like ⋙ 12.

Then you could list statements with ⋙ in between to show where one statement ends and the other begins.

∃xBx⋙(P ∨ Q)⋙Axyz

Elegant: Who needs ⋙?

If you mash together various sentences of L, you can always tell where one sentence stops and the next begins.

Let's take the example given above without the separator symbol:

∃xBx(P ∨ Q)Axyz.

There is always a `mechanical` way to separate them into sentences.

The algorithm:

You start.

If you hit a quantifier and a variable, keep going

You will then either hit a predicate letter or a "("

If it's a predicate letter, go till you hit either a quantifier or a (

If it's a "(", then stop after you have encounter an equal number of "(" and ")"

Then using either the **quick and dirty** or the *Elegant* approach, you can turn a finite list of statements that represent a proof into a single Gödel number.

Chapter Eleven
Doing Proofs by the Numbers

Fred felt a real sense of power in working in the arithmetic model. He had defined **SSSS**1 as 5, so the constants were much easier to write. 487 was much easier than writing **SSSSSSSSSSS** . . . **SSS**1 with 486 **S**s.

2 + 3 was much easier to write than A**S**1,**SS**1.

And inside that arithmetic model, he could compute the Gödel number of any sentence $\mathcal{S}$. He could go backwards and take any Gödel number and find the sentence $\mathcal{S}$ that generated that number.

He could take an proof (a list of sentences) and compute its Gödel number. He could go from that Gödel number back to the proof.

In the model he defined the 2-ary predicate proves(**a**, **b**) to be true iff there is a proof (whose Gödel number is **b**) that proves the sentence (whose Gödel number is **a**).

Sometimes he would say "**b** proves **a**," as a shorthand for "the proof that **b** represents proves the sentence that **a** represents."

Everyone knows that a number can't prove another number.

If he were mean—which he isn't—he would assign as homework the question: "Is it true that proves(5674981531, 2315646874896488000)?"

The student would first have to find the sentence whose Gödel number is 5674981531.

Then the student would have to find the list of sentences whose Gödel number is 2315646874896488000.

Then determine the reason for each line of the proof.

Then check whether the last line of the proof is the sentence that we're trying to prove.

Some notes . . .

♪#1: The object isn't to be able to go through all the arithmetic in converting Gödel numbers into sentence(s). The whole point is to note that this *can be done.*

♪#2: Every sentence $\mathcal{S}$ in the arithmetic model of L can be converted into its Gödel number, but most natural numbers are *not* Gödel numbers of *sentences*. If I take some random number like 26487700 it can be converted into $2^{n_1}3^{n_2}5^{n_3}7^{n_4}$ . . . where the n_i are natural numbers. But that might represent ♦♦P∃∀)1, which isn't a sentence in the model.

After defining proves(a, b), he would define the 1-ary predicate

provable(a) iff there is a constant b such that proves(a, b).

Let's put these two definitions in a box so that they will be easy to find.

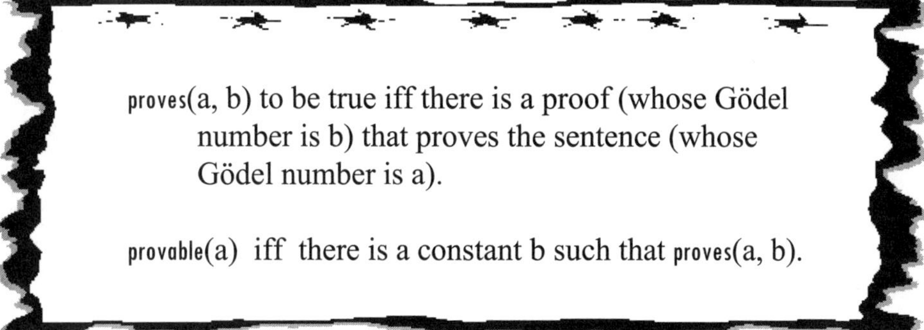

proves(a, b) to be true iff there is a proof (whose Gödel number is b) that proves the sentence (whose Gödel number is a).

provable(a) iff there is a constant b such that proves(a, b).

There is a World of Difference between saying "b proves a"—proves(a, b)—and saying "a can be proved"—provable(a).

When you assert proves(3253515361000, 2516761801268810000), you can `mechanically` check to see whether it is true. Translation: Fred could build a machine into which you insert 3253515361000 and 2516761801268810000, and it will tell you either **T** or **F**.

Translation: The 2-ary predicate proves is decidable.

Translation: You could write a computer program into which you could input any two natural numbers, and the program would print out either ☺ or ☹. It might take a long time to run, but you could be sure that eventually you would get one of those two outputs.

In contrast, saying "a can be proved"—provable(a)—is not decidable. Translation: No computer program will ever be written that is able to tell whether provable(a) is true or false for *every* natural number a.
Translation: In 1931 Gödel's Incompleteness Theorem showed we will never find an algorithm that will always work in determining whether a particular sentence $\mathcal{S}$ in the arithmetic model is true.

I, your reader, don't believe all this poppycock. * ***If Fred can build a machine that can take*** proves(a, b) ***for any*** a ***and*** b ***and output either*** T ***or*** F, ***then I can write a program that can output either*** T ***or*** F ***for*** provable(a) ***for any*** a.

Please tell me more. If you can do that, then Gödel's 1931 Incompleteness Theorem is wrong. You will become famous.

THE KITTEN Caboodle

The Official Campus Newspaper of KITTENS University Saturday 6 p.m. Edition 10¢

latebreaking news

Gödel's Theorem Overturned!

KANSAS: Minutes ago, an intrepid reader of *Life of Fred: Logic* has just proved that all the fame and glory that Kurt Gödel has received by proving his 1931 Incompleteness Theorem was all a mistake.

This evening at 9 p.m. this reader will unveil the computer program that will take any statement $\mathcal{S}$ in the arithmetic model and determine whether it is true.

The presentation will be held in Archimedes Hall. Cookies and Sluice will be available after the talk.

Wait a minute! I may not be able to write that program in three hours. And give a talk! I'd prefer just to explain it to you, Mr. Author, and you can do the talk.

Okay. It's a deal. But you make the cookies. Tell me, how can you turn a program that can check whether any proof of a sentence $\mathcal{S}$ in

*nonsense or foolishness. *Poppycock* entered the language in 1840–1850.

the arithmetic model is correct—proves(a, b)—into a program that can tell whether any sentence 𝒮 is true—provable(a)?

It's easy peasy.** *I'll go slow so you can understand it.

Thank you.

First, you are given any old sentence 𝒮. It will have a Gödel number, say, 4568000. I need to find out whether 𝒮 is true.

I first run the program proves(4568000, 1). ***If it outputs* T*, I'm done. I have found a proof.***

If not, then I run proves(4568000, 2). ***If it outputs* T*, then I've found a proof of*** 4568000 ***and I'm done.***

Otherwise, I run proves(4568000, 3). ***And so on.***

I have a small thought. What if 𝒮 happens to be false? Then your program will run forever and never determine the truth value of 𝒮.

I got that figured out. Let's suppose that the Gödel number of ¬𝒮 ***is*** 7554000.

I first run proves(4568000, 1) ***and then run*** proves(7554000, 1).

Then I run proves(4568000, 2) ***and then run*** proves(7554000, 2).

Then I run proves(4568000, 3) ***and then run*** proves(7554000, 3) ***and so on until one of the two gives me a* T.**

Either 𝒮 or ¬𝒮 has gotta be true. Right?

That's right.

So this new program must eventually show that either 𝒮 or ¬𝒮 is true. Right?

Nope.

What do you mean "Nope"!

Your program might never stop. You are assuming that it will.

In Chapter 5 ("Stinky Logic") we talked about Fallacy #6: Reasoning in a circle. You have just done that.

Please explain.

You are trying to show that every sentence 𝒮 is decidable. You are trying to show that for every 𝒮, either 𝒮 or ¬𝒮 must be provable.

---

*Actually, the full expression is easy peasy lemon squeezy. The phrase originated in the United Kingdom (1976). It's much better than the American equivalent: *easy as pie* because *easy peasy lemon squeezy* rhymes.

You wrote a program that will check any known proof (with Gödel numbers 1, 2, 3, 4, 5 . . .) for either $\mathcal{S}$ or $\neg\mathcal{S}$. Then you claimed that the program will eventually find a proof for either $\mathcal{S}$ or $\neg\mathcal{S}$. But that claim that it will stop is *exactly what you were trying to show in the first place.*

Oops. Cancel the cookies.

Our goal is to prove Gödel's Incompleteness Theorem.

Translation: Our goal is to find a sentence $\mathcal{G}$ in the arithmetic model of L that is both true and unprovable. When you input $\mathcal{G}$ into your program, it will never stop.

Translation: After we find $\mathcal{G}$ we must *prove* that $\mathcal{G}$ is both true and unprovable.

Skip this box if you have a perfect memory.

In constructing the arithmetic model from L . . .

1. In logic language L we identified one constant and called it 1. (one)

2. In the arithmetic model we defined a 1-ary predicate **S** (the successor function) and gave five Peano postulates describing the properties of **S** and 1.

3. We defined 2, 3, 4, 5 . . . as **S**1, **S S**1, **S S S**1, **S S S S**1. . . .

4. In five days of Fred's lecturing (in *Life of Fred: Five Days*) he developed all of arithmetic—addition, subtraction, multiplication, division, exponents, fractions. . . . *How* he did that is not critical for understanding Gödel's Incompleteness Theorem. It was just a lot of fun showing that Peano's Postulates could do so much.

5. The constants can be thought of as the natural numbers 1, 2, 3, 4. . . .

6. We could take any sentence $\mathcal{S}$ in the arithmetic model and find its Gödel number.

A sentence $\mathcal{S}$ in logic must be either **T** or **F**.

Some random string of symbols in L might not be either **T** or **F**. For example, (((P $\vee$ & $\Rightarrow$ **S**abc((P is just pure nonsense.

Some strings are *almost* sentences. If $\mathcal{S}$ is a sentence, every variable must be bound. (See pages 80–81 for definition of *bound*.) For example, 6 + x = 17 is neither **T** nor **F**, but if we put a quantifier in front of it, it becomes a sentence.

$\forall$x (6 + x = 17) is **F**.

$\exists$x (6 + x = 17) is **T** .

Quick Fred! I need a particular natural number. What's your favorite one?

I have lots of them.

Just give me one. I want to use it for the rest of the book.

Okay. How about 8?

You are weird. Most people say either 7 or 1. Okay. We'll go with 8.

In the arithmetic model, we can take any sentence $\mathcal{S}$ and replace every 8 with some other constant, say, b.

For example, $\forall x(8^2 = 55x)$ would become $\forall x(b^2 = 55x)$.

It's done by looking at the Gödel numbering of $\forall x(8^2 = 55x)$ and locating any 8s and replacing them with the Gödel number corresponding to b.

We will symbolize this by subst($\mathcal{S}$, 8, b).*

Since we are going to keep Fred's favorite 8 throughout the rest of the book, we can abbreviate subst($\mathcal{S}$, 8, b) by $\mathcal{S}$(b).

$\mathcal{S}$(b) means taking sentence $\mathcal{S}$ and stuffing a "b" everywhere you see an "8."

$\mathcal{S}$(b) is a new sentence in L, unless, of course, $\mathcal{S}$ didn't have any 8s in it to begin with. Without any 8s, $\mathcal{S}$(b) would just remain $\mathcal{S}$.

All we need to prove Gödel's Incompleteness Theorem is one lemma: the Diagonal lemma. (Fancy people call it the Diagonalization lemma.) Once we've proven that lemma, the proof of the theorem is duck soup.**

---

*subst is short for substitution.

** = easy peasy. Duck soup entered our language in 1910–1915.

Chapter Twelve
The Diagonal Lemma

Prologue: This is the hardest part of the book. Please fasten your seatbelt. It's going to take about ten lines to prove this lemma. No logic books at the high school level prove this. Virtually no freshman college logic books prove this. The proof is usually offered at the graduate level (after the bachelor's degree).

When Fred presented the Diagonal lemma in his class, he first put the statement of the lemma on the board, and then he talked about it for a while before he proved it.

The **Diagonal lemma**:

For any sentence $\mathscr{S}$
there is a sentence $\mathscr{T}$ with Gödel number t,
such that we can prove
$\mathscr{T} \Leftrightarrow \mathscr{S}(t)$.

Fred's preliminary discussion:

We start with a sentence $\mathscr{S}$.

$\mathscr{S}$(t) means, by definition, subst($\mathscr{S}$, 8, t).

This means that you substitute t for every 8 that you can find in sentence $\mathscr{S}$.

There are two possibilities.
Either $\mathscr{S}$ contains 8s or it doesn't.

If it doesn't, then $\mathscr{S}$(t) is the same as $\mathscr{S}$.
If $\mathscr{S}$ is **T** then set $\mathscr{T}$ equal to 1+1 = 2, and $\mathscr{T} \Leftrightarrow \mathscr{S}(t)$.
If $\mathscr{S}$ is **F** then set $\mathscr{T}$ equal to 1+1 = 3, and $\mathscr{T} \Leftrightarrow \mathscr{S}(t)$.

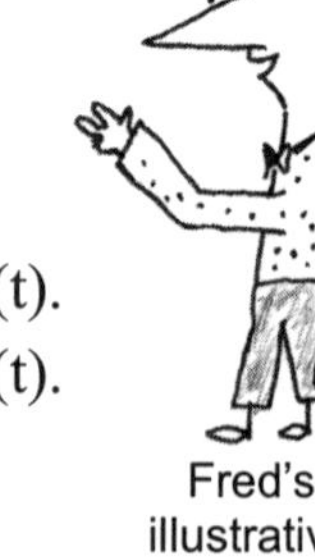
Fred's illustrative gesture

If $\mathscr{S}$ does contain 8s, then we have some work to do.

We are given a sentence $\mathcal{S}$. Let me just try out some various sentences $\mathcal{T}$ and see if one of them will accidentally do the job.

Suppose $\mathcal{T}$ is **T** and the Gödel number for $\mathcal{T}$ is 5413.
Checking, I find that $\mathcal{S}(5413)$ is **F**. That didn't work.
Suppose $\mathcal{T}$ is **T** and the Gödel number for $\mathcal{T}$ is 74420.
Checking, I find that $\mathcal{S}(74420)$ is **F**. That didn't work.

Suppose $\mathcal{T}$ is **F** and the Gödel number for $\mathcal{T}$ is 900.
Checking, I find that $\mathcal{S}(900)$ is **T**. That didn't work.
Suppose $\mathcal{T}$ is **F** and the Gödel number for $\mathcal{T}$ is 5840066770000.
Checking, I find that $\mathcal{S}(5840066770000)$ is **T**. That didn't work.

Rats! I'll never find the $\mathcal{T}$ just by trying numbers at random.

PROOF OF THE DIAGONAL LEMMA

First, we notice that any string of symbols in L can have a Gödel number. It doesn't need to be a sentence (that is either **T** or **F**). Nowhere in Chapter 10 in defining a Gödel number did we use the fact that the string of symbols is a sentence.

Just take the string of symbols, such as xP), and use the chart

↓ ( ) ∀ ∃ **S** 1 P c x ♦
1 2 3 4 5 6 7 8 9 10 11 to translate xP) into 10, 8, 3 and then stick them as the exponents $2^{10}\,3^{8}\,5^{3}$ and we have its Gödel number.

Second, we need an abbreviation to make the proof of the Diagonal lemma shorter and easier to read. diag(s) = b means:

1. Find the string of symbols S whose Gödel number is s.
2. Then determine the Gödel number of the string of symbols $\exists x(x = s \,\&\, S)$. That number is b.

In English diag(s) can be read as, "Take the diagonal of the string that has Gödel number s."

Third, there is no third. We are ready to do the proof.

The proof . . .

Start with any sentence $\mathscr{S}$ in the arithmetic model of L.

Recall that $\mathscr{S}$(y) was defined as subst($\mathscr{S}$, 8, y).

1. Let $\mathscr{C}$ be the string ∃y(diag(x) = y & $\mathscr{S}$(y)).
2. Let c be the Gödel number of $\mathscr{C}$.

Please slow way down! You won't understand anything if you read this proof like you read a novel. An English major will fill the trunk of a car with all the novels, essays, plays, etc. that are required reading. They have to read at a mile a minute in order to get through all those wordy-birdy books. English majors have to use eye drops in order to prevent their eyes from squeaking.

A math major can carry four years of required math reading in a briefcase or two.

Expect this proof, which has less than a dozen lines, to take a full hour to absorb.*

3. Let $\mathscr{T}$ be the sentence ∃x(x = c & $\mathscr{C}$).
4. Let t be the Gödel number of $\mathscr{T}$.

Puzzle #47: How could I call $\mathscr{T}$ a sentence, when $\mathscr{C}$ was only a string of symbols?

5. $\mathscr{T}$ = ∃x(x = c & ∃y(diag(x) = y & $\mathscr{S}$(y)))

Puzzle #97: How did I get line 5.?

We're half way there. Only five more lines to go.

6. $\mathscr{T}$ ⇔ ∃y∃x(x = c & diag(x) = y & $\mathscr{S}$(y))

---

*Fred is super bright. It only took him 59 minutes.

Lines 5 and 6 look almost alike. I just moved the ∃y all the way to the left. I used the tautology ∃x(P & ∃yQ) ⇔ ∃y∃x(P & Q) in the case in which P does not contain the variable y.

To prove this tautology I would have to establish ⇔, which means that I would have to show both ∃x(P & ∃yQ) ⇒ ∃y∃x(P & Q) and its converse ∃y∃x(P & Q) ⇒ ∃x(P & ∃yQ).

The proof would begin

a. ∃x(P & ∃yQ)	premise
b. subst(P, x, c) & ∃y subst(Q, x, c)	∃ elim where c is a constant not in line a.
c. subst(P, x, c)	taut from line b
d. ∃y subst(Q, x, c)	taut from line b
e. subst(subst(Q, x, c), y, d)	∃ elim where d is a constant not in any previous line
f. subst(P, x, c) & subst(subst(Q, x, c), y, d)	taut from lines c and e
g. ∃x(P & subst(Q, y, d))	∃ intro from line f
h. ∃y∃x(P & Q)	∃ intro from line g

1. ∃x(P & ∃yQ) ⇒ ∃y∃x(P & Q) from the lemma

i. ∃y∃x(P & Q)	premise
j. etc.	

You get the idea. I don't want to spend a whole lot of time with this old stuff. Let's get back to the proof of the Diagonal lemma.

Repeating line 6 so you won't have to turn back to the previous page:

6. $\mathcal{T}$ ⇔ ∃y∃x(x = c & diag(x) = y & $\mathcal{S}$(y))

7. $\mathcal{T}$ ⇔ ∃y∃x(diag(c) = y & $\mathcal{S}$(y))

Puzzle #43: What is the reason for line 7?

8. $\mathcal{T}$ ⇔ ∃y∃x(t = y & $\mathcal{S}$(y))

Puzzle #77: diag(c) was replaced by t. Why is this true?

9.	$\mathcal{T} \Leftrightarrow \exists y \exists x \mathcal{S}(t)$	by the same reason we got line 7, namely by the substitution property of equality.
10.	$\mathcal{T} \Leftrightarrow \mathcal{S}(t)$	since there are no unbound y's or x's in $\mathcal{S}(t)$

We did it! Q.E.D. We showed that for any sentence $\mathcal{S}$, we could find a sentence $\mathcal{T}$ with Gödel number t, such that $\mathcal{T} \Leftrightarrow \mathcal{S}(t)$.

You may unfasten your seatbelt now. We've done the hard part. The proof of Gödel's Incompleteness Theorem (also known as Gödel's First Incompleteness Theorem) will be a snap.*

---

*To be a snap = easy peasy.

Chapter Thirteen
Gödel's Incompleteness Theorem

In 1931 Gödel proved his Incompleteness Theorem. He became the most celebrated logician in recent centuries. This theorem is considered by many mathematicians to be the most central (and surprising) part of logic. We are going to prove it in this chapter.

Gödel's Incompleteness Theorem:* In the model of arithmetic there is a sentence that is true and is not provable.

Proof . . .

Let $\mathcal{S}$ be provable(a).**

No. I take that back. Let $\mathcal{S}$ be ¬provable(a).

By the Diagonal lemma there is a sentence $\mathcal{G}$ with Gödel number g such that $\mathcal{G} \Leftrightarrow$ ¬provable(g). Q.E.D.

Wait! Hold it. That's it? That was terribly short. Could you please add a couple of extra words? You got to* Q.E.D. *too quickly.

I'd be glad to, but there isn't much to say.

What does $\mathcal{G} \Leftrightarrow$ ¬provable(g) mean? $\mathcal{G}$ and ¬provable(g) have the same truth value (by the meaning of $\Leftrightarrow$).

That means that either $\mathcal{G}$ and ¬provable(g) are both **T**, or they are both **F**.

Case 1: They are both **T**. Then we are done since we have found a sentence $\mathcal{G}$ that is both true and unprovable.

Case 2: $\mathcal{G}$ and ¬provable(g) are both **F**, but this is impossible.

---

*Also known as Gödel's First Incompleteness Theorem. We call it the first theorem, because there is a second one. The second one is important, but not as earthshaking as the first.

**Remember the box on page 125?

provable(a) meant that there is a constant b such that proves(a, b).

proves(a, b) to be true iff there is a proof (whose Gödel number is b) that proves the sentence (whose Gödel number is a).

If ¬ provable(g) is **F**,

that means that provable(g) is **T**.

That means that $\mathscr{G}$ is **T**.

But in this Case 2, we know that $\mathscr{G}$ is **F**.

Gödel's Incompleteness Theorem takes 16 words to state:

In the model of arithmetic there is a sentence that is true and is not provable.

It has a two-sentence proof:

Let $\mathscr{S}$ be ¬ provable(a).

By the Diagonal lemma there is a sentence $\mathscr{G}$ with Gödel number g such that $\mathscr{G} \Leftrightarrow$ ¬ provable(g).

You are a math major. You call up your English major friends and invite them to join you for pizza and a movie tonight. They say no, because they have to read some 400-page novel and write a paper that is due tomorrow.

You call up your history major friends, and they say no, because they have to read history books (with dates and names of kings and words like *extraterritoriality**) and write a paper.

And music majors and ballet majors have to practice day and night. The competition in those fields is intense. The old expression: "Many are majored, but few are hired."

---

*Extraterritoriality means that cops can't give parking tickets to foreign diplomats.

Chapter Fourteen
Reader Revolts!

I can't believe it. I, your reader, have been thinking about your Gödel's Incompleteness Theorem. The whole thing seems stupid, and I am hereby taking over the writing of this logic book. I need the space of this Chapter 14 to express two very big objections that I have.

I like the idea that math majors don't have the slavery of writing long papers, but this Gödel stuff seems plain goofy.

Oh?

This is my chapter and I want you to hear me out before you try to make me feel better.

Okay.

I'll be the lawyer and ask you, Mr. Author, the questions I have. I will also be the judge, so that any objections you have will be overruled.

Geep.

I move that Geep be stricken from the record. ~~Geep.~~

The witness will just answer my questions.

I read your proof of Gödel's Incompleteness Theorem, which stated that there is a sentence $\mathcal{G}$ ***in the arithmetic model with the properties: (1)*** $\mathcal{G}$ ***is true and (2)*** $\mathcal{G}$ ***is not provable. I couldn't find anything wrong with your proof, but the conclusion is nuts. My question: Did you prove that*** $\mathcal{G}$ ***is true?***

Yes I did. I proved that $\mathcal{G}$ is true.

My second question: Did you prove that $\mathcal{G}$ ***is not provable?***

Yes I did. I proved that $\neg$provable(g) where g is the Gödel number of $\mathcal{G}$.

Answer my question! Did you prove that $\mathcal{G}$ ***is not provable?***

Gulp. Yes. I proved that $\mathcal{G}$ is not provable in the arithmetic model.

Guilty! Do you have anything to say before I pass sentence and condemn this *Life of Fred: Logic* ***book to the garbage can?***

Yes.

Say it.

If the court will walk with me out to the Courtyard of Mathematics I have something I need to show to you.

I could use some fresh air. Let's go.

Here are the statues (the structures, the models) celebrating the many subfields (the theories) of mathematics.

At the bottom of each structure are the axioms from which the tree would grow. In the arithmetic model, we find the five Peano postulates.

The ground of the Courtyard of Mathematics is paved with LOGIC. Each of the structures rests on that bedrock.

Are you just killing time? I know all of that already.

As we walk around the Courtyard, let's stand in front of the arithmetic model.

When I was *in* the model, I showed how to take any string of symbols of L and, using arithmetic, find its Gödel number.

Okay. So what?

And I showed how to find the Gödel number of any list of sentences using the `quick and dirty` method.

That was Chapter 10 stuff.

Then in the arithmetic model I defined proves(a, b) when the proof whose Gödel number was b would prove the sentence whose Gödel number was a.

Then defined in the model provable(a) iff there is a constant b such that proves(a, b).

That was Chapter 11 stuff. Please get to the point.

I am! Everything I had done up to the Diagonal lemma was done *in* the arithmetic model. I was hanging on that palm tree making definitions and proving theorems.

Then I climbed off the tree and stood on the ground in the Courtyard of Mathematics and looked at the arithmetic structure.

I don't know what you're talking about.

Patience please. This is bad news that I'm trying to break to you s l o w l y.

When I got to Chapter 12, I proved the Diagonal lemma: For any sentence $\mathcal{S}$, there is a sentence $\mathcal{T}$ with Gödel number t, such that $\mathcal{T} \Leftrightarrow \mathcal{S}(t)$.

I didn't prove that *in* the arithmetic model! I proved the Diagonal lemma while standing on the ground of the Courtyard of Mathematics looking *at* the arithmetic structure.

The Diagonal lemma was proved *outside* the arithmetic model. Are you ready?

I'm ready.

The Diagonal lemma begins, "For any sentence $\mathcal{S}$. . .". In symbols that would be $\forall\mathcal{S}$. In the model the only symbol that follows $\forall$ can be a variable like x.

So I did prove that $\mathcal{G}$ was true, and I did prove ¬provable(g), but ¬provable(g) means not provable *in* the arithmetic model. There is no contradiction.

You are tricky. You didn't tell me you were doing math while standing on the pavement of the Courtyard looking at the model.

I knew the truth would come out eventually. I just didn't want to pile extra stress on you while we were proving the Diagonal lemma.

There is a name for all of this. When you are standing on the ground and looking at a theory and figuring out which theorems can or can't be proven when you are in the model, this is called **metamathematics**.

Not all universities offer courses in metamathematics. In several of the earlier *Life of Fred* books you have seen Fred's teaching schedule:

Fred Gauss
—room 314—

8–9 Beginning Algebra
9–10 Advanced Algebra
10–11 Geometry
11–noon Trigonometry
noon–1 Calculus
1–2 Statistics
2–3 Linear Algebra
3–3:05 Break
3:05–5 Seminar in Biology, Economics, Physics, Set Theory, Topology, and Metamathematics.

Now you know what means.

Not all universities are high-powered enough to offer mathematics this deep. KITTENS University offers it. The University of California at

Berkeley offers Math 225A and Math 225B—two graduate courses in metamathematics. Each is a 4-unit course. In the courses descriptions they mention "predicate logic" and "completeness theorems" and "theory of models" and "undecidable theories." In this *Life of Fred: Logic* book that you are holding in your hands—and are *not* going to throw into the garbage can—we have played with predicate logic, with the completeness of first-order logic*, with all the models in the Courtyard of Mathematics, and with our proof of the undecidable model of arithmetic (not every true sentence can be proved).

Okay. You are acquitted of the charge of contradicting yourself when you said that 𝒢 and ¬provable(g) are both true.

But I have a second charge against you, Mr. Author. This is something that will really spin the propeller on your cap.

You claim that there is a sentence 𝒢 with the property that 𝒢 and ¬provable(g) are both true. I'll go along with that.

Now suppose I take your arithmetic statue with its five Peano postulates and add 𝒢 to that list. We now have six axioms. Certainly 𝒢 is true. Now I'm going to make it an axiom.

But all your talk about ¬provable(g) being true is just hogwash.** I can prove 𝒢 easily.

Puzzle #15: Prove 𝒢.

Ha, Ha, Ha, Ha, Ha, Ha, Ha, Ha, Ha, Ha, Ha, Ha, and Ha!

Is that your argument? Looking back on the list of 17 popular fallacies in reasoning (page 70), I would rate your "Ha, Ha, Ha, Ha, . . . and Ha!" as #3 (verbal abuse).

---

*Recall Kurt Gödel's 1929 doctoral thesis with the snappy title "Die Vollständigkeit der Axiome des logischen Funktionenkalküls" with its 12-page proof of the completeness of logic.

**= nonsense

But the thought of sticking $\mathcal{G}$ as one of the axioms in the arithmetic structure (model, statue) is a clever idea. It certainly will be true that I could no longer say that $\mathcal{G}$ is not provable in the model.

Thank you. I knew I was right.

However, there is one little thing that should be pointed out. When you change the axioms, you change the theory. You change which statue that we are looking at. You are erecting a new statue in the Courtyard of Mathematics.

Hey. If I'm making a new theory (a new statue), then I don't want that wimpy palm tree stuff. I want a real structure.

Reader's New Model

Granted.

And $\mathcal{G}$ will be one of the theorems of you new structure.

And in your new structure, all of the Gödel numbering still works since you haven't changed the symbols of L.

And in this reader's new model, we now have proofs that depend on the six axioms (the Peano postulates plus $\mathcal{G}$).

And in your new statue I can define a 2-ary predicate proves(a, b) to be true iff there is a proof (whose Gödel number is b) that proves the sentence (whose Gödel number is a). And I can define provable(a) iff there is a constant b such that proves(a, b).

These proves and provable are not the same ones we had in the original arithmetic model. Stuff, such as $\mathcal{G}$, are now provable.

Now I repeat all of Chapter 12 (the Diagonal lemma) and all of Chapter 13 (Gödel's Incompleteness Theorem) and get a new **G** for this structure. This new **G** is both true and not provable in your new reader's model.

I know what I'll do. I'll stick both $\mathcal{G}$ and G as axioms. Oh, forget it. I just realized what you would do.

Puzzle #85: What would I do?

Have I answered your objections?

Yes. Thank you. You have been patient with me. I turn the writing of this book back to you.

Chapter Fifteen
Arithmetic Model vs. Set Theory Model

Fred and everyone else likes axiom systems (models) that are consistent, complete, and meaningful.

consistent = can't prove false statements such as P & ¬P
complete = can prove every true statement
meaningful = applies to the real world

We know that arithmetic is super useful. It's meaningful.

Gödel's Incompleteness Theorem says that the arithmetic model is not complete.

We have mentioned two important structures (set theory in Chapter 8 and arithmetic theory in Chapters 9–14). Since any structure that is "infected" with arithmetic is not complete, one obvious question is:

Is set theory complete?

Good old set theory. ❦ ∈ {✈, ❦, ❀, ⌚}
$\{\pi, R\} \cup \{R, \xi\} = \{\pi, R, \xi\}$
$\{3, 4, 5, 9\} \cap \{2, 3, 5\} = \{3, 5\}$

The axioms of set theory are the eight Zermelo-Fraenkel axioms, some of which were mentioned in Chapter 8. (All of which are discussed in detail in *Life of Fred: Five Days.*)

The answer is no. Set theory is not complete.

Set theory is at least as "strong" as arithmetic. Translation: Everything that can be done in arithmetic can also be done in set theory.

In arithmetic we had two special symbols: 1 and **S** (the successor function). In set theory we will represent 1 by the empty set Ø (= { }) and represent **S**c (where c is any set) by {c, {c}}.

So, for example, 2 = **S**1 = { Ø, { Ø } }. All of the Peano postulates will hold. Set theory is incomplete.

Chapter Sixteen
Consistent

It was 6:30 p.m. on Saturday. In the crepuscular light, Fred walked around the Courtyard of Mathematics. When he looked at the statues that represented the theories of arithmetic and of set theory, he thought *They are stunted. They can't prove all the true things about arithmetic or set theory. And, of course, since virtually all the statues depend on arithmetic or set theory, they are all "undertall."*

At three feet tall, Fred could identify with all those math theories.

Then he spotted one huge, gigantic, reach-to-the-skies, cloud-capped, statue. It was tall.

It was loaded with theorems. It had proofs of $2 + 2 = 4$,

of the isosceles triangle theorem,

of the Goldbach conjectures,

of . . .

Before you, my reader, yell your "I object!" I do remember that the truth of the Goldbach conjectures that I mentioned on page 19 is not known today. But those conjectures of Goldbach are proven theorems in this super-tall math theory.

This theory also has proofs . . .
of everything that Mr. Duck has ever said, and of

of $6 + 7 = 11$, and

of sugar is good for you.

In short, this theory is inconsistent—you can prove *any* statement is true.

It's very easy to build one of these prove-everything theories. All you need is rotten axioms at its base.

All it takes is one rotten axiom such as P & ¬P and the statue grows into outer space.

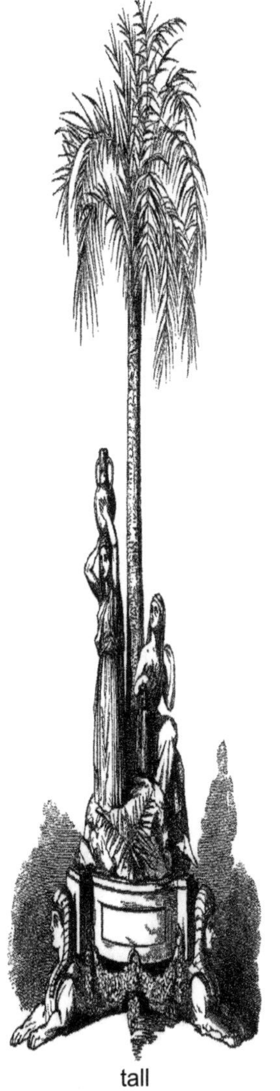
tall

That is because if you start with untruth, everything that follows is twaddle.*

☞ For example, people who think that they are the most important thing in the world (an obvious untruth!) will often make the silliest pronouncements.**

☞ For example, people who think that they are absolutely worthless and insignificant (an obvious untruth!) will often mumble things like, "It's okay if you step on my face. I'm a nobody."

A theory is consistent if you can only prove true statements.

A theory is complete if every true statement can be proved.

Consistent & Complete are brother and sister.

Gödel's First Incompleteness Theorem said that in the arithmetic (or set theory) model, there are true statements that can't be proved in the model. The theories are not complete. Translation: stunted statues.

What about consistency? In the arithmetic model, can we prove that we can only prove true statements? Translation: Can we prove that we'll never prove P & ¬P? Translation: Can we prove in the arithmetic model that hidden deep inside the five Peano postulates there isn't some contradiction lurking?

Gödel's Second Incompleteness Theorem says no. We will never be able to prove *inside the model* (while we are on the tree) that this tree won't grow to the sky.

Of course, Gödel proves that metamathematically. (I always wanted an opportunity to say that octosyllabic word.) He proves it looking at the statue.

The proof is short. When we proved the Diagonal lemma, that was the hard part of a course in logic. Both Gödel's First and Second Incompleteness Theorems have short proofs.

---

*= poppycock = foolishness. Logic books need lots of words to describe the illogical.

**If you think you are the King or Queen of the Universe, you don't just make statements. You make pronouncements.

Before I begin the proof, I want to note that "contradictions can prove anything" was one of the tautologies in our famous list on page 55. The penultimate* tautology was $\models (R \;\&\; \neg R) \Rightarrow Q$. You proved this as Puzzle #63 on page 92.

Now to the work of proving that we will never be able to prove a contradiction in the theory of arithmetic.

The definition of consistent is ¬provable(R & ¬R).

Let's abbreviate ¬provable(R & ¬R) as Const.

little lemma

First note that if P and Q are any two statements with Gödel numbers p and q, then $P \Rightarrow Q$ will imply that provable(p) ⇒ provable(q). Here is why. Suppose I know $P \Rightarrow Q$ and I know provable(p). The definition of provable(p) means that there is a proof of P.

That means that there is a list of statements in the theory where the last statement is P.

Attach to that list one more line. That line will be Q with the reason $P \Rightarrow Q$ and *modus ponens*.** That will be a proof of Q.

The proof of Const . . .

We know that $\mathscr{G}$ is true	Gödel's First Incompleteness Theorem
Const ⇒ $\mathscr{G}$	$\models P \Rightarrow (Q \Rightarrow P)$ 7th taut on page 55
provable(Const) ⇒ provable(g)	the little lemma
¬provable(g)	Gödel's First Incompleteness Theorem
¬provable(Const)	$\models (\neg Q \;\&\; (P \Rightarrow Q) \Rightarrow \neg P)$ 2nd taut on page 55 (*modus tollens*).

Q.E.D.

*= next-to-last

** $\models (P \;\&\; (P \Rightarrow Q)) \Rightarrow Q$

$$\begin{array}{l} P \\ \underline{P \Rightarrow Q} \\ Q \end{array}$$

Fred sat on a bench in the Courtyard of Mathematics and gazed at the arithmetic statue. His feet didn't touch the ground, but he was used to that.

This looked like a wonderful week of logic lectures coming up. On Tuesday and Wednesday he would present propositional and predicate logic.

Thursday would be the tough day: Gödel numbering and the ten-line proof of the Diagonal lemma. Friday would be easy: Gödel's First and Second Incompleteness Theorems. Both proofs are very short.

the metamathematics bench

For the proof of the First Theorem, the Diagonal lemma implied that there is a sentence $\mathcal{G}$ with Gödel number g such that $\mathcal{G} \Leftrightarrow \neg$ provable(g). Then there were two cases: Either $\mathcal{G}$ and $\neg$ provable(g) are both **T**, or they are both **F**. If they were both **T**, then the proof was over. If they were both **F**, then that leads to the mess of $\mathcal{G}$ being both **T** and **F**.

The Second Theorem was just a corollary of the First. The First showed that $\neg$ provable(g) was true. Translation: There was a sentence in the arithmetic model that could not be proven in the model. That automatically showed that the arithmetic model was consistent, because in any inconsistent theory, you can prove everything.

Gödel's Second Incompleteness Theorem showed that consistency was not provable *inside* any model containing the Peano postulates. Fred knew this while sitting on the metamathematics bench looking at the arithmetic statue. He was happy, happy, happy.

And then it clicked.

He looked over at one of those ultra-tall inconsistent statues (theories)—one that reached to the sky and could prove everything. The particular one he looked at just had the Peano postulates and 1 = 2 as the axioms.

By the Second Incompleteness you couldn't prove consistency inside the model. But you could, since you could prove anything!

His whole week of teaching would be ruined. He would be 35 minutes into his Friday lecture. He would have finished the proofs of the two theorems, and Betty would raise her hand and point out the fact that inconsistent statues were consistent.

Somewhere in the proofs of the Diagonal lemma and the two theorems he had messed up.

Now comes the one of the hardest puzzles in this book. Only 0.037% of readers will be able to solve it in under five minutes. In a day's thinking 0.8% of readers will see where Fred went wrong.

If you spend a week or a year working on this problem, it will be time well spent. I, your author, spent 14 years working on a geometry problem before I solved it. It is one of the joys of my life.

Puzzle #99: Gödel's Second Incompleteness Theorem is obviously false.

That theorem states that we will never be able to prove inside any model that contains the Peano postulates that the model is consistent.

Translation: If a model contains the Peano postulates, you can't prove consistency. But that is obviously false.

It's false because the model having as axioms the Peano postulates and R & ¬R can easily prove anything, including consistency.

Where is the mistake?

Fred took three seconds and found the mistake. He hopped off the bench and headed into the campus library. He was feeling good again.

Puzzle #1: Perhaps the easiest thing is to ask Duck, "Do you know what the capital of Kansas is?"

If he says, "Yes," then you know that he doesn't know.

If he says, "No," then you are happy. He knows!

small essay

About Just Looking at the Question and Reading My Answer

Learning the facts about logic is only a part of the purpose of *Life of Fred: Logic.*

A long time ago, you spent about a year studying high school geometry. If it was a decent course, most of your effort involved doing proofs of geometry theorems. One easy part of the course was the definitions. An isosceles triangle is a triangle with at least two congruent sides. Another easy part was seeing that the theorems were true. The base angles of an isosceles triangle are congruent.

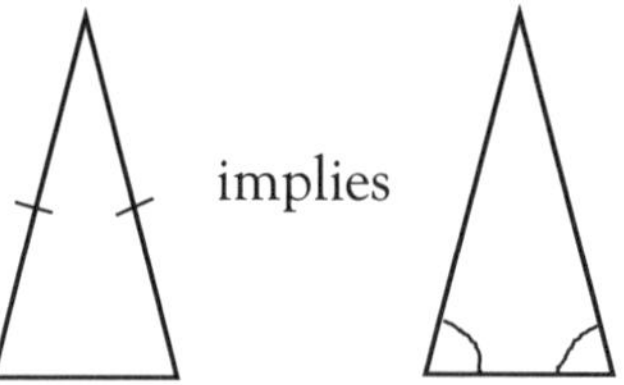

The tough part was proving those theorems. Did you ever ask yourself (or the teacher) why you were busting your brain doing proofs? Everyone knows that thinking is the hardest work there is. Brain surgeons come home and go and dig in the garden and plant roses. Do professional gardeners come home and do brain surgery?

Learning "stuff"—such as the answers to the logic puzzles or how to flip a burger or empty a bed pan—may get you a minimum wage job.

Exercising your brain—such as doing geometry proofs or solving logic puzzles—makes your brain able to learn more quickly. It's not knowing a lot of "stuff" that's important. Anyone with access to the internet has access to zillions of facts.

Your success in almost any high-paying job will depend on your ability to learn fast while on the job.

end of small essay

Puzzle #4: Which of these are true?

A) $\exists x \neg Sx$

This translates in the model as, "There is someone who is not a student at KITTENS." We know that is true since Sd is false.

B) $\forall x \neg Sx$

This translates in the model as, "Everybody is not a student at KITTENS." This isn't true since Sb.

C) $\neg \forall x\ Sx$

This translates as, "It is not the case that everyone is a student at KITTENS." This is true since Sd is false.

Wait a minute! A) and C) say the same thing. $\exists x \neg Sx$ says the same thing as $\neg \forall x\ Sx$. To say that there exists a case in which Sx isn't true is to say that it's not always true that Sx.

Someday (maybe in the next chapter) we will prove that

$$\exists x \neg Ax \Rightarrow \neg \forall x\ Ax$$

D) $\neg \exists x\ Sx$

This translates as, "It is not the case that there exists someone who is a student at KITTENS." That is false because Sb.

And you may guess what's coming next. We note that B) and D) say the same thing. $\forall x \neg Sx$ says the same thing as $\neg \exists x\ Sx$. Both say that no one is a student at KITTENS. $\forall x \neg Sx \Leftrightarrow \neg \exists x\ Sx$

Puzzle #5: What tautology did I use to get line e?

In other words how did I get from $Ms \Rightarrow Ds$ and Ms to Ds?

In other words how did I get from $P \Rightarrow Q$ and P to Q?

The answer is good old *modus ponens.* I told you we would use that propositional tautology frequently.

Puzzle #6: Finish up these last two line of the proof.

From the proof we had . . .

a. $\exists x \neg Ax \ \&\ \forall x\ Ax$	premise
. . .	
g. $\neg \forall x\ Ax$	taut from lines e and f

1. $(\exists x \neg Ax \ \&\ \forall x\ Ax) \Rightarrow \neg \forall x\ Ax$	from the lemma
2. $\exists x \neg Ax \Rightarrow \neg \forall x\ Ax$	taut from line 1

Puzzle #7: Using the recursive definition of a sentence in L, show that $\neg\neg\neg B$ is a sentence in L.

B is a sentence in L by 1. Every sentence letter is a **sentence** in L.
$\neg B$ is a sentence in L by 2. If P is a sentence in L, then so is $\neg P$.
$\neg\neg B$ is a sentence in L by 2. If P is a sentence in L, then so is $\neg P$.
$\neg\neg\neg B$ is a sentence in L by 2. If P is a sentence in L, then so is $\neg P$.

Puzzle #8: $\vdash \exists x(Ax \vee Bx) \Rightarrow \exists x\, Ax \vee \exists x\, Bx$

Proof . . .

a.	$\exists x(Ax \vee Bx)\ \&\ \neg(\exists x\, Ax \vee \exists x\, Bx)$	premise
b.	$\exists x(Ax \vee Bx)$	taut from line a
c.	$\neg(\exists x\, Ax \vee \exists x\, Bx)$	taut from line a
d.	$\neg\exists x\, Ax\ \&\ \neg\exists x\, Bx$	taut from line c one of De Morgan's laws
e.	$\neg\exists x\, Ax$	taut from line d
f.	$\neg\exists x\, Bx$	taut from line d
g.	$Ac \vee Bc$	$\exists$ elim from line b

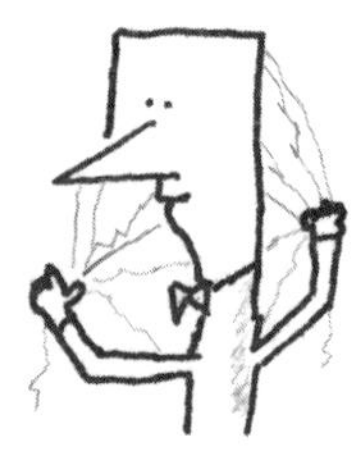

Now here is where things get a little sticky.

Lines e, f, and g can't all be true. There is a contradiction between those lines.

But I can't use a $\exists$ elim on line e to get $\neg Ac$. That's because constant c has already been used on line g.

Lines e and f would be a lot nicer if they had $\forall$s instead of $\exists$s.

h.	$\forall x\, \neg Ax$	taut from line e replacing $\neg\exists$ with $\forall\neg$
i.	$\forall x\, \neg Bx$	taut from line f
j.	$\neg Ac$	$\forall$ elim from line h
k.	$\neg Bc$	$\forall$ elim from line i

The challenge is now to show that lines g, j, and k can't coexist happily.

l.	$\neg Ac\ \&\ \neg Bc$	taut from lines j and k
m.	$\neg(Ac \vee Bc)$	taut from line l De Morgan strikes again.
o.	etc.	taut from lines g and m

Puzzle #10: In Boolean algebra, let P → i. Your challenge is to find ¬P in terms of i.

Your answer may be different than mine.
At first, I tried i – 1. That would take 1 to 0 and 0 to –1.
Then I thought, How can I keep the 0 and turn the –1 into a 1?
My final answer: $(i - 1)^2$.

Puzzle #11: Write ZF#3 ∀C∀D∃A(C ∊ A & D ∊ A) in the language of L.

∀x∀y∃z(Mxz & Myz)
(Your choice of variables may be different than mine.)

Puzzle #15: Prove *𝒢*.

We are starting with six axioms—the five Peano postulates and the reader's sixth postulate:

Peano postulate #1: 1 is a natural number.
Peano postulate #2: For every natural number n, 𝔖n is also a natural number.
Peano postulate #3: 1 is not the successor of any natural number.
Peano postulate #4: No two different numbers have the same successor.
Peano postulate #5: Suppose you have a set K and you know two things:
(i) 1 is in K and
(ii) for every natural number n, if n in is K, then 𝔖n is also in K,
then K is the set of all natural numbers.
Reader's #6: *𝒢*

Proof of *𝒢*. . .

Statement	Reason	
1. *𝒢*	Postulate #6.	Q.E.D.

Puzzle #18: [harder] Argue from the definition of ⇒, which has this truth table

	P	Q	P ⇒ Q
line 1 →	**T**	**T**	**T**
line 2 →	**T**	**F**	**F**
line 3 →	**F**	**T**	**T**
line 4 →	**F**	**F**	**T**

that A ⇒ B and B ⇒ A, must mean that A ⇔ B.

Argument: If A $\Rightarrow$ B is true, then we must be dealing with lines 1, 3, or 4 in the truth table.

If B $\Rightarrow$ A, then we must be dealing with lines 1, 2, or 4 in the truth table.

Since we are given both A $\Rightarrow$ B and B $\Rightarrow$ A, we must be dealing with either line 1 or line 4 in the truth table.

Line 1 says both A and B are true.
Line 4 says both A and B are false.

A and B are logically equivalent, which is the definition of A $\Leftrightarrow$ B. (See the definition of $\Leftrightarrow$ on page 33.)

Puzzle #25: Sentence letters like P are really predicate letters! If P = "Portia is pretentious," we could say that P is a _____ predicate.

If A = " . . . is ascorbic" and 1-ary
if B = ". . . runs faster than . . . and 2-ary
if C = ". . . becomes . . . in the presence of . . ." 3-ary

It wouldn't be too surprising to say that sentence letters like P is a **0-ary predicate**.

Our language L used to contain seven items:

1) sentence letters
2) connectives
3) parentheses
4) predicate letters
5) constants
6) variables
7) quantifiers.

We have combined 1) and 4).

L now has six items:

1) predicate letters (0-ary, 1-ary, 2-ary, . . .)
2) connectives
3) parentheses
4) constants

5) variables
6) quantifiers.

Language Lesson:

0-ary predicates are sometimes called sentence letters or zero-place predicates.

1-ary predicates are sometimes called one-place predicates or monadic predicates.

2-ary predicates are sometimes called two-place predicates or binary predicates or dyadic predicates.

3-ary predicates are sometimes called three-place predicates or trianary predicates or triadic predicates.

It's from *binary* and *trianary* that the suffix *-ary* was developed for 0-ary, 1-ary, 2-ary, etc.

Puzzle #30: List the other three things (for the other three connectives) that have to be true for function h.

In addition to 1. h(¬P) = **T** iff h(P) = **F**.
2. h(P & Q) = **T** iff h(P) = **T** and h(P) = **T**.

we would have 3. h(P ∨ Q) = **T** iff h(P) = **T** or h(Q) = **T** or both.
4. h(P ⇒ Q) = **T** iff h(P) = **F** or h(Q) = **T** or both.
5. h(P ⇔ Q) = **T** iff h(P) = h(Q).

Puzzle #31: Translate Peano Postulate #4: No two different numbers have the same successor.

$\forall x \forall y(f(x) = f(y) \Rightarrow x = y)$ or you might have written

$\forall x \forall y(\neg Exy \Rightarrow \neg(f(x) = f(y))$, which is the contrapositive.

Both are correct.

Puzzle #32: Logicians say which of these (if any) are true?

A) If the moon is made of cheese, then mice made the craters.

This is the case in which **F** ⇒ **F**. Therefore, the sentence is true.

B) If 2 + 2 = 4, then 785 = 785.

Here is the easiest case. **T** ⇒ **T** is a true sentence.

C) If Elvis was the first president of the U.S. then King Tut was left handed.

You have no idea whether King Tut was left handed. Neither do I. But that isn't important. What we do know is that Elvis wasn't the first president, so we have the situation **F** ⇒ ?. This is always true.

D) If King Tut was left handed, then Washington was the first president of the U.S.

Again, we don't know if King Tut was left handed. We are in the situation ? ⇒ **T**. Looking at the truth table for implies, it isn't difficult to see that if Q is **T**, then P ⇒ Q will always be **T**.

Puzzle #33: Prove ⊢ ∀x Ax ⇒ Ac.

		a. ∀x Ax	premise
		b. Ac	∀ elim
1. ∀x Ax ⇒ Ac	from the lemma		

Puzzle #34: Find a model in which $((\forall x(Px \Rightarrow Qx) \& \neg Pc) \Rightarrow \neg Qc$ is not valid.

I bet your model will be different than mine! I will let P be the 1-ary predicate "___ is a pizza meat topping mentioned in *Life of Fred: Dreams*"* and Q is the 1-ary predicate "___ can be found at the PieOne pizza restaurant and x is any object.

Then $\forall x(Px \Rightarrow Qx)$ asserts that any pizza meat topping can be found at PieOne. This is true.

If I let c be a measuring cup, then $\neg Pc$ asserts that a measuring cup is not a pizza meat topping. This is true.

Qc asserts that a measuring cup can be found at the PieOne restaurant. This is true. Then $\neg Qc$ is false.

Putting all these together, $((\forall x(Px \Rightarrow Qx) \& \neg Pc) \Rightarrow \neg Qc$ becomes (**T** & **T**) ⇒ **F**, which is false.

Puzzle #35: Your question: Is this a model for L?

A model for L must assign a sentence to each sentence letter in L. A sentence is defined in logic to be a statement that is either true or false. A sentence cannot be an opinion (Strawberry is the best ice cream flavor), a question (Do you carry strawberry ice cream?), a command (Eat it!), or an interjection (Yummy!).

All these three sentences are either **T** or **F**:

☞ "Next Tuesday a 100-megaton atomic bomb will be detonated over KITTENS University."
☞ "Some people at KITTENS will never reach their 60th birthday."
☞ "Some people think carrots are orange."

Therefore, this is a model for L.

---

*which includes: bacon, beef, chicken, chorizo, duck, honey-cured ham, meatballs, pepperoni, salami, sausage, turkey, venison, American bison, steak, veal, yak, hare, rabbit, kangaroo, opossum, goat, ibex, pork, reindeer, moose, antelope, giraffe, squirrel, whale, bear, duck, dove, quail, mutton, ostrich, pheasant, grouse, partridge, pigeon, woodcock, goose, frog, anchovy, crab, cod, eel, halibut, salmon, shark, abalone, clam, conch, oyster, scallop, crayfish, lobster, prawns, salmon, shrimp, squid, and tuna.

Life of Fred: Dreams is the fourth book in the Language Arts series for high school English.

Puzzle #37: (easy) What if we need 24 constants? How do we avoid running into the land reserved for variables?

We use the same trick as we did to get an infinite number of sentence letters (P_1, P_2, P_3, . . .) or an infinite number of variables (x_1, x_2, x_3, . . .).

We may never need a_1 or a_2 or a_{48}, but it's nice to know that they are there should an emergency arise.

Puzzle #38: Establish $\vDash (P \& (P \Rightarrow Q)) \Rightarrow Q$.

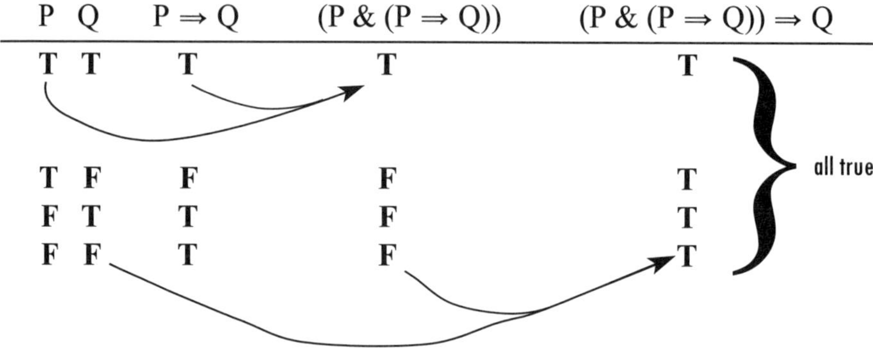

P	Q	P ⇒ Q	(P & (P ⇒ Q))	(P & (P ⇒ Q)) ⇒ Q
T	**T**	**T**	**T**	**T**
T	**F**	**F**	**F**	**T**
F	**T**	**T**	**F**	**T**
F	**F**	**T**	**F**	**T**

Puzzle #39: (a little harder) What tautology did I use to get line d from line a?

In other words how did I get from $\forall x(Mx \Rightarrow Dx) \& Ms$ to Ms?
In other words how did I get from P & Q to Q?

In other words how did I get $(P \& Q) \Rightarrow Q$?
I'll do the truth table for you.

P	Q	P & Q	(P & Q) ⇒ Q
T	**T**	**T**	**T**
T	**F**	**F**	**T**
F	**T**	**F**	**T**
F	**F**	**F**	**T**

Or I could have dug down into the list of popular tautologies on page 55 and found the Law of Simplification, which is the tenth one on the list.

Puzzle #40: Create an *arithmetic* model for L.
Your answer will be different than mine.
Assign A to ¼ + ⅔ = 27.
Assign B to 5 – 3 = 2.
Assign C to 6 × 0 = 3,232,534,898.007
Assign D to π/π = 1.
Assign all other letters to 4 = 4.

Puzzle #42: If you have n sentence letters, how many possible models would you need?

1 ➞ 2
2 ➞ 4
3 ➞ 8 8 is 2^3
4 ➞ 16 16 is 2^4
n ➞ 2^n

Puzzle #43: What is the reason for line 7?

In line 6 I knew that x = c, so I substituted c in place of x in the expression diag(x).

Puzzle #44: List the 16 lines needed to list all the possible **T**s and **F**s when dealing with four letters.

T T T T
T T T F
T T F T
T T F F
T F T T
T F T F
T F F T
T F F F
F T T T
F T T F
F T F T
F T F F
F F T T
F F T F
F F F T
F F F F

There are zillion ways to make this list (and not lose your mind).

One way is to list the right column first. It just has **T**s and **F**s alternating: **T F T F** etc.
Then list the second-from-the right column. It just has pairs of **T**s and **F**s alternating: **T T F F T T** .
Then list the third-from-the right column. It just has **T T T T F F F F T T T T F F F F**.
The left column has 8 **T**s followed by 8 **F**s.

A second way to write out this list is a little like a recursive definition. For one sentence you have **T**
F .

To get two sentence letters you take the **T**
F
and write it twice.

T
F
T
F and stick a **T** in front of the first half and an **F** in front of the second half.

T T
T F
F T
F F .

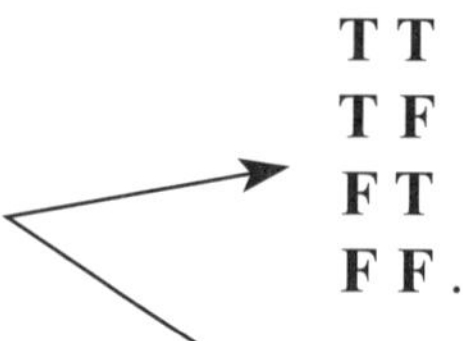

For three sentence letters you take [the above] and write it twice and put a **T** in front of the first half and an **F** in front of the second half.

Each time you add a new letter, you double the number of rows.

A third way to write out a list of **T**s and **F**s is to start with as many **T**s are there are sentence letters. In this Puzzle #44, we start with 4 **T**s.

T T T T . Now think of **T**s as ones and **F**s as zeros.

We now have 1 1 1 1.

For the next row list the number that is just barely smaller than 1 1 1 1.

We now have 1 1 1 1
1 1 1 0.

Now list the number just barely smaller than 1 1 1 0 *that doesn't have any 2s, 3s, 4s, 5s, 6s, 7s, 8s, or 9s in it.*

Now we have	1	1	1	1	
	1	1	1	0	
	1	1	0	1	We skipped 1109, 1108, 1107, 1106, 1105, 1104, 1103, and 1102.
Then . . .	1	1	0	0	
Then . . .	1	0	1	1	Skipping 1099, 1098, . . . , 1012.
All the way down to	0	0	0	0	

Finally, you convert all these 1s and 0s back to **T**s and **F**s. This third way seems the most complicated, but if you are a robot or you are working with computers, this convert-to-numbers approach is the easiest to implement.

Puzzle #45: ⊢ ∃x∀y Axy ⇒ ∀y∃x Axy

Proof . . .		
	a. ∃x∀y Axy & ¬∀y∃x Axy	premise
	b. ∃x∀y Axy	taut from line a
	c. ¬∀y∃x Axy	taut from line a
	d. ∃y¬∃x Axy	taut from line c replacing ¬∀ with ∃¬
	e. ∃y∀x ¬Axy	taut from line d replacing ¬∃ with ∀¬

In proofs, it's often helpful to do the ∃ elim steps before the ∀ elim steps. That's because the requirement for doing ∃ elim is that the constant introduced must be new to the proof.

In contrast, when you do a ∀ elim you can introduce *any* constant.

In this proof, I'm going to first "blast away" the ∃s from lines b and e. Then I'll "blast away" the ∀s.

f. $\forall$y Aby	$\exists$ elim from line b The constant b is new to the proof.
g. $\forall$x ¬Axc	$\exists$ elim from line e The constant c is new to the proof.
h. Abc	$\forall$ elim from line f
i. ¬Abc	$\forall$ elim from line g
j. etc.	taut from lines h and i

Puzzle #46: Write the Disjunctive syllogism in this artistic mode with a line separating the premises from the conclusion.

$\models ((P \vee Q) \,\&\, \neg P) \Rightarrow Q$ becomes

$$\frac{\begin{array}{c} P \vee Q \\ \neg P \end{array}}{Q}$$

My Dad says that on our vacation we are either going to Phoenix (capital of Arizona) or we are going to Quincy (in Illinois*). Later he says that we aren't going to Phoenix. The Disjunctive syllogism tells me

that . . .

Puzzle #47: How could I call $\mathscr{T}$ a sentence, when $\mathscr{C}$ was only a string of symbols?

$\mathscr{C}$, which was $\exists$y(diag(x) = y & $\mathscr{S}$(y)), has an x in it that was not bound. Sentences in L have every variable bound.

$\mathscr{T}$, which was $\exists$x(x = c & $\mathscr{C}$), bound that variable x.

Puzzle #50: Using this recursive definition, establish that ((A ∨ B) ⇒ (C & B)) is a sentence in L.

❶ A, B, and C are sentences in L by 1. Every sentence letter is a **sentence** in L.

❷ (A ∨ B) is a sentence in L by 4. If P and Q are sentences in L, then so is (P ∨ Q).

❸ (C & B) is a sentence in L by 3. If P and Q are sentences in L, then so is (P & Q).

❹ ((A ∨ B) ⇒ (C & B)) is a sentence in L by 5. If P and Q are sentences in L, then so is (P ⇒ Q).

---

*Springfield is the the capital of Illinois.

Puzzle #51: Which of these are true?
A) ∃x Px Since Glen is playing, Pg is true, and therefore ∃x Px is true.
B) ∀x Px Not everyone is playing since Pf is false. ∀x Px is false.

C) ∀x(Px ⇒ Hx)

This is true. It's always the case that if you are playing tennis, you are holding a racquet.

D) ∀x(Hx ⇒ Px)

This is false. It's is not always the case that if you are holding a racquet, you are playing tennis. Some people just use a tennis racquet as an air guitar.

E) ∃x(Hx ⇒ Px)

This will be true if we can find at least one value of x for which Hx ⇒ Px. It is certainly true for Hg ⇒ Pg, because Glen is both holding a tennis racquet and is playing. **T** ⇒ **T** is true.

It happens to be also true for Hf ⇒ Pf, because both Hf and Pf are both false. **F** ⇒ **F** is true by our definition of ⇒.

Puzzle #52: ⊢ ∀x(Ax ⇒ Bx) ⇒ (∃x Ax ⇒ ∃x Bx)

This is my solution. Yours may be different.

Proof . . .	a. ∀x(Ax ⇒ Bx) & ¬(∃x Ax ⇒ ∃x Bx)	premise
	b. ∀x(Ax ⇒ Bx)	taut from line a
	c. ¬(∃x Ax ⇒ ∃x Bx)	taut from line a
	d. Ac ⇒ Bc	∀ elim from line b
	e. ∃x Ax ⇒ Bc	∃ intro from line d
	f. ∃x Ax ⇒ ∃x Bx	∃ intro from line e
	g. etc.	

Oh yes. I forgot. You like me to write out the last three mechanical steps. Okay. I'll do it this time.

g. (∃x Ax ⇒ ∃x Bx) — taut from lines c and f

1. (∀x(Ax ⇒ Bx) & ¬(∃x Ax ⇒ ∃x Bx)) ⇒ (∃x Ax ⇒ ∃x Bx) — from lemma
2. ∀x(Ax ⇒ Bx) ⇒ (∃x Ax ⇒ ∃x Bx) — taut from line 1

Puzzle #55: We have P & Q → ij. Now all we need is the math equivalent for ¬P and we are done. Why is that?

All five Boolean connectives (¬, &, ∨, ⇒, and ⇔) can be expressed just in terms of ¬ and &. Back on page 40, we said that ¬and & are expressively complete.

That means that, for example, P ∨ Q can be written as ¬(¬P & ¬Q). So all we need are the Boolean algebra expressions for ¬ and &. We are half way home, because we already have P & Q → ij.

Puzzle #56: How many different ways in English could you express "If Kingie paints a picture, then someone will buy it."

1. Every picture Kingie paints will be sold.
2. Kingie painting a picture implies that it will be sold.
3. Kingie painting a picture is a sufficient condition for it to be sold.
4. If the picture didn't sell, then Kingie didn't paint it.
5. All of Kingie's paintings sell.
6. For every painting in the world it is true that either Kingie didn't paint it or that it sold (or both).
7. For a painting to sell, it is sufficient to know that Kingie painted it.
8. Being sold is a necessary result of Kingie painting a picture.
9. All pictures painted by Kingie will be purchased.

Do you remember Venn diagrams from algebra days? All of these nine sentences can be expressed by

Puzzle #60: Find a model in which ∃x Ax ⇒ ∀x Ax is not valid.

Your answer will probably be different than mine. In fact, I'm sure it will be different. In my structure the 1-ary predicate A will mean "____is the oil painting that Kingie created that was entitled *Autumnal Antics*."* In my structure the constants will be all oil paintings.

"Autumnal Antics"
by Kingie

∃x Ax is true because *Autumnal Antics* is one of Kingie's most famous paintings.

∀x Ax is false because not every oil painting is Kingie's *Autumnal Antics.*

Puzzle #62: Using only ¬, &, and ∨, express A ⇒ B.

Just before this puzzle question, I wrote, "A ⇒ B will be true if and only if either A is false or B is true (or both)."

Either A is false or B is true (or both) is ¬A ∨ B.

So A ⇒ B iff ¬A ∨ B.

Puzzle #63: Prove ⊨ (R & ¬R) ⇒ Q.

R	Q	¬R	R & ¬R	(R & ¬R) ⇒ Q
T	**T**	**F**	**F**	**T**
T	**F**	**F**	**F**	**T**
F	**T**	**T**	**F**	**T**
F	**F**	**T**	**F**	**T**

**Autumnal* is the adjective form of *autumn*, or more correctly, it is the adjectival form of *autumn*.

Puzzle #65: Establish ⊨ (P & Q) ⇔ ((P ↓ P) ↓ (Q ↓ Q))

P	Q	P & Q	P↓P	Q↓Q	(P↓P)↓(Q↓Q)	(P & Q) ⇔ ((P↓P)↓(Q↓Q))
T	**T**	**T**	**F**	**F**	**T**	**T**
T	**F**	**F**	**F**	**T**	**F**	**T**
F	**T**	**F**	**T**	**F**	**F**	**T**
F	**F**	**F**	**T**	**T**	**F**	**T**

If you just looked at the puzzle and then turned to this answer without writing down your answer, then you would have just seen this boring truth table. Instead, if you had worked this puzzle for yourself, you would have spent several minutes and probably made several errors, but you would have learned a lot more. The idea of what P ↓ P means would be clearer. Your trading your time/energy/pain/effort for knowledge is the deal that nature offers you.

When I was in high school, we studied myths from older cultures. I still remember one of them—it's been over a half century since I heard the myth—about some guy who found a fountain whose waters could impart all kinds of knowledge and wisdom. The keeper of the fountain told the guy that in order to drink he would have to leave one of his eyes in the fountain. He made the exchange.

If you really want to learn logic, all I'm asking for is a minute or two of your effort. You can keep your eyeballs.

Puzzle #70: How many possible logical connectives could be placed between two sentences in logic?

There are two possibilities for the truth of P ✱ Q in the case that P is true and Q is true. P ✱ Q could be either **T** or **F**.

In the second case (P true, Q false), P ✱ Q could be either **T** or **F**.
In the third case (P false, Q true), P ✱ Q could be either **T** or **F**.
In the fourth case (P false, Q false), P ✱ Q could be either **T** or **F**.

By the fundamental principal of counting, there are 2×2×2×2 possible connectives. The happy news is that we won't need all 16 connectives. In fact, we won't need all four connectives—&, ∨, ⇒, and ⇔. Would you care to make a guess: How many is the minimum number of connectives needed to do all the logic stuff? We'll see in the next chapter. If you can't wait till then, here's the Duck to give you a hint.

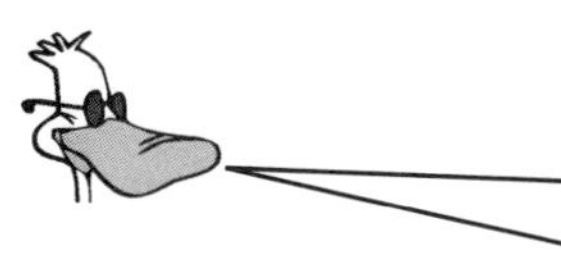

The minimum number of connectives is either 2 or 3 or 4 or 5 or 6 or 7 or 8 or 9 or 10 or 11 or 12 or 13 or 14 or 15 or 16.

Puzzle #71: $\vdash (\exists x\ Ax \vee \exists x\ Bx) \Rightarrow \exists x(Ax \vee Bx)$

Proof . . .

a. $\exists x\ Ax\ \&\ \neg\exists x(Ax \vee Bx)$ — premise

Note that I'm only using the $\exists x\ Ax$

b. $\exists x\ Ax$ — taut from line a

c. $\neg\exists x(Ax \vee Bx)$ — taut from line a

d. Ac — $\exists$ elim from line b

We are at the same sticking point as we were in the previous proof (puzzle #8). Lines c and d can't both be true. There is a contradiction between those two lines.

But you can't use $\exists$ elim on line c to get $\neg(Ac \vee Bc)$ because the constant c was already used on line d.

Line c would be a lot nicer if it had $\forall$ instead of $\exists$.

e. $\forall x\ \neg(Ax \vee Bx)$ — taut from line c

replacing $\neg\exists$ with $\forall\neg$

f. $\forall x(\neg Ax\ \&\ \neg Bx)$ — taut from line e

De Morgan

g. $\neg Ac\ \&\ \neg Bc$ — $\forall$ elim from line f

h. $\neg Ac$ — taut from line g

i. $\exists x(Ax \vee Bx)$ — taut from lines d and h

1. $(\exists x\ Ax\ \&\ \neg\exists x(Ax \vee Bx)) \Rightarrow \exists x(Ax \vee Bx)$ — from the lemma
2. $\exists x\ Ax \Rightarrow \exists x(Ax \vee Bx)$ — taut from line 1

We are half way done. This was the first case.

Now we do it for the second case $\exists x\ Bx \Rightarrow \exists x(Ax \vee Bx)$.

j. $\exists x\ Bx\ \&\ \neg\exists x(Ax \vee Bx)$ — premise

k. etc. Here we repeat lines b through i making the necessary changes (= *mutatis mutandis*).

3. $(\exists x\ Bx\ \ \&\ \neg\exists x(Ax \lor Bx)) \Rightarrow \exists x(Ax \lor Bx)$ from the second lemma
4. $\exists x\ Bx \Rightarrow\ \exists x(Ax \lor Bx)$ taut from line 3

Having done both cases, we have only one more line in the proof.

5. $(\exists x\ Ax \lor \exists x\ Bx) \Rightarrow \exists x(Ax \lor Bx)$ taut from lines 2 and 4

Using the tautology $\models ((P \Rightarrow R)\ \&\ (Q \Rightarrow R)) \Rightarrow ((P \lor Q) \Rightarrow R)$

Puzzle #75: In a single sentence explain why P is a sentence letter and why Pc is a predicate letter followed by a constant.

Attempt ❶: *Because sentence letters don't have constants coming after them.* ⇦ fail. This is a sentence fragment.

Attempt ❷: *everyone can see that predicate letters are the ones that have a constant attached to them.* ⇦ fail. Sentences begin with a capital letter.

Attempt ❸: *Sentence letters do not have a constant right after them. Predicate letters do have a constant right after them.* ⇦ fail. I asked for a *single* sentence.

Attempt ❹: *Sentence letters do not have a constant right after them, predicate letters do have a constant right after them.* ⇦ fail. In English classes this is called a **comma splice** or a **run-on sentence**.

Attempt ❺: *Without going into too much detail and within the confines of a single sentence, one would differentiate between sentence letters and predicate letters by noting that all predicate letters are concatenated with a constant.* ⇦ This is a single English sentence—which I asked for—but stylistically it stinks. The use of stilted language—such as "one would differentiate—and ten-dollar words such as *concatenated*, when "followed by" would suffice, make this a nightmare for readers. In addition it is prolix.

Puzzle #76: $\vdash (\forall x(Ax \Rightarrow Bx)\ \&\ \forall x(Bx \Rightarrow Cx))\ \Rightarrow\ \forall x(Ax \Rightarrow Cx)$

Proof . . .

a. $(\forall x(Ax \Rightarrow Bx)\ \&\ \forall x(Bx \Rightarrow Cx))\ \&\ \neg\forall x(Ax \Rightarrow Cx)$ premise

b. $\forall x(Ax \Rightarrow Bx)\ \&\ \forall x(Bx \Rightarrow Cx)$ taut from line a

c. $\neg\forall x(Ax \Rightarrow Cx)$ taut from line a

Did I mention that the big difference would be a lot more writing?

d. $\exists x\ \neg(Ax \Rightarrow Cx)$ taut from line c

replacing $\neg\forall$ with $\exists\neg$

e. $\neg(\text{Ad} \Rightarrow \text{Cd})$	$\exists$ elim from line d
f. $\forall x(\text{A}x \Rightarrow \text{B}x)$	taut from line b
g. $\forall x(\text{B}x \Rightarrow \text{C}x)$	taut from line b
h. $\text{Ad} \Rightarrow \text{Bd}$	$\forall$ elim from line f
i. $\text{Bd} \Rightarrow \text{Cd}$	$\forall$ elim from line g
j. $\text{Ad} \Rightarrow \text{Cd}$	taut from lines h and i
k. etc.	taut from lines e and j

Okay. I'll write out the last three mechanical lines.

k. $\forall x(\text{A}x \Rightarrow \text{C}x)$ taut from lines e and j

1. $((\forall x(\text{A}x \Rightarrow \text{B}x)\ \&\ \forall x(\text{B}x \Rightarrow \text{C}x))\ \&\ \neg\forall x(\text{A}x \Rightarrow \text{C}x)) \Rightarrow \forall x(\text{A}x \Rightarrow \text{C}x)$
from the lemma

2. $(\forall x(\text{A}x \Rightarrow \text{B}x)\ \&\ \forall x(\text{B}x \Rightarrow \text{C}x)) \Rightarrow \forall x(\text{A}x \Rightarrow \text{C}x)$
taut from line 1

Puzzle #77: diag(c) was obviously replaced by t. Why is this true?

The definition of diag(c) is the Gödel number of $\exists x(x = c\ \&\ \mathscr{C})$ where c is the Gödel number of $\mathscr{C}$.

$\exists x(x = c\ \&\ \mathscr{C})$ is $\mathscr{T}$ by line 3.

The Gödel number of $\mathscr{T}$ is t by line 4.

Puzzle #78: Which of these is true? $P \Rightarrow Q$, $Q \Rightarrow P$, and $P \Leftrightarrow Q$?

If you square any even natural number, you will get an even answer. Translation: $P \Rightarrow Q$ is true.

If you start with any perfect square natural number that is even, such as 144, and you take the square root, you will get an even answer. Translation: $Q \Rightarrow P$.

Once you have both $P \Rightarrow Q$ and $Q \Rightarrow P$, you automatically have $P \Leftrightarrow Q$ by Puzzle #18.

Puzzle #84: Draw a Venn diagram to illustrate $\neg\exists x(\text{A}x\ \&\ \text{B}x)$.

This says that there isn't an x that makes both Ax and Bx true.

In the language of set theory, we would say that the sets A and B are disjoint.

In a Venn diagram . . .

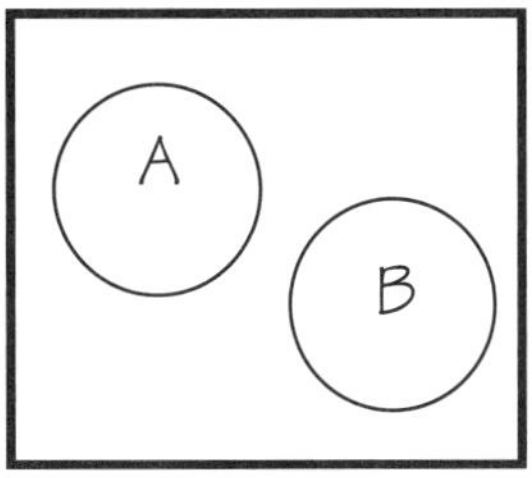

Puzzle #85: What would I do if we make both $\mathscr{G}$ and 𝔾 as new axioms?

If you do that, then you have a new statue (a new theory) and I repeat Chapters 12 and 13 (Diagonal lemma and Gödel's Incompleteness Theorem) in this new environment and create a new $\mathcal{G}$ that is both true and unprovable.

I need to mention, before I forget, that the sentence G that is produced by the Diagonal lemma and Gödel's Theorem is named after Gödel. Everyone calls it G (or $\mathscr{G}$ or 𝔾 or $\mathcal{G}$). It's that sentence that Gödel showed was both true and unprovable. Someday, you, my reader, may have a sentence named after you. There are 25 letters left in the alphabet.

Puzzle #88: There may be more than one way to have Duck admit that Topeka is the capital of Kansas. One way is to ask him, "What *isn't* the capital of Kansas?" The only way that he can lie is to tell you that Topeka isn't the capital.

Puzzle #90: $\vdash \neg\forall x\forall y\exists z\, Axyz \Rightarrow \exists x\exists y\forall z \neg Axyz$

Proof . . .		
	a. $\neg\forall x\forall y\exists z\, Axyz$	premise
	b. $\exists x \neg\forall y\exists z\, Axyz$	taut from line a
	c. $\exists x\exists y \neg\exists z\, Axyz$	taut from line b
	d. $\exists x\exists y\forall z \neg Axyz$	taut from line c

1. $\neg\forall x\forall y\exists z\, Axyz \Rightarrow \exists x\exists y\forall z \neg Axyz$ from the lemma

Mechanical #92:

Part A Is it true that $\vDash (A \Leftrightarrow B) \Leftrightarrow \neg(A \Leftrightarrow \neg B)$?

A	B	$A \Leftrightarrow B$	$\neg B$	$A \Leftrightarrow \neg B$	$\neg(A \Leftrightarrow \neg B)$	$(A \Leftrightarrow B) \Leftrightarrow \neg(A \Leftrightarrow \neg B)$
T	**T**	**T**	**F**	**F**	**T**	**T**
T	**F**	**F**	**T**	**T**	**F**	**T**
F	**T**	**F**	**F**	**T**	**F**	**T**
F	**F**	**T**	**T**	**F**	**T**	**T**

Since the last column is all **T**s, this is a tautology.

Part B Is it true that $\models A \Rightarrow (B \Rightarrow A)$?

A	B	$B \Rightarrow A$	$A \Rightarrow (B \Rightarrow A)$	
T	**T**	**T**	**T**	
T	**F**	**T**	**T**	
F	**T**	**F**	**T**	
F	**F**	**T**	**T**	A tautology.

Part C Is it true that $\models (A \Rightarrow B) \Rightarrow A$?

A	B	$A \Rightarrow B$	$(A \Rightarrow B) \Rightarrow A$	
T	**T**	**T**	**T**	
T	**F**	**F**	**T**	
F	**T**	**T**	**F**	We stop at this point. It is not a tautology.

Puzzle #93: Continuing Puzzle #35, which, if any, of these are true: $H \Rightarrow C$, $C \Rightarrow H$, or $H \Leftrightarrow C$?

If you detonate that giant bomb over KITTENS no one will survive. In particular, Fred, 6, will never reach his 60th birthday. Therefore, $H \Rightarrow C$ is true.

It is true that some people at KITTENS University will never reach their 60th birthday. In any population of thousands of students, it is virtual certain that at least one of them will die before reaching 60. So C is true.

On the other hand, H is false. You know that and I know that. I have complete authorial control over what happens here at KITTENS and at the current rate* Fred will be around longer than any of my grandkids. In this model, $C \Rightarrow H$ is **T** ⇒ **F**. This implication is false.

$H \Leftrightarrow C$ is not true since $H \Rightarrow C$ is true and $C \Rightarrow H$ is false.

Sudden thought: Considering $A \Rightarrow B$, $B \Rightarrow A$, and $A \Leftrightarrow B$,

① They can't all be false.
② Exactly one can be true. (That was this Puzzle #93.)
③ Exactly two of them can't be true.
④ All three of them can be true.

Only an odd number of $A \Rightarrow B$, $B \Rightarrow A$, and $A \Leftrightarrow B$ can be true. No other logic textbook (that I know of) has ever noticed this.

---

*This is my 55th book and Fred is 6.

Puzzle #94: Which of the 17 fallacies apply?

There may be some legitimate disagreement with the answers that I have chosen. Sometimes more than one fallacy may apply. Consider these answers as my *suggestions*.

Argument A: #15: Hasty Generalization. Appealing to inductive logic with too few examples.

Argument B: #13: Red Herring. Finding other areas of agreement that have nothing to do with the conclusion that is being sought. OR #11: *Non Sequitur.* The premises don't support the conclusion.

Argument C: #14: Appeal to Questionable Authority.

Argument D: #9: Appeal to Pity.

Argument E: #3: Verbal Abuse.

Argument F: #2: Appeal to Ego—the listener's desire for respect, love, or security.

Argument G: #10: The Straw Man. Set up a fake argument and knock it down. A more important reason than the dislike of your wife might be the fact that you are already married.

Puzzle #95: Why is $\forall x \exists y(Px \ \&\ Qy) \Rightarrow Px$ *not* a sentence in predicate logic?

In the last three pages of Chapter 6, we talked about the scope of a quantifier and that a variable y is bound iff it is in the scope of $\exists y$ or $\forall y$. We noted that in every sentence we make in first-order logic, *every variable is bound.*

In the expression $\forall x \exists y(Px \ \&\ Qy) \Rightarrow Px$, that last x isn't bound.

Puzzle #96: Give a half dozen models for ∀x Px ⇒ ∃x Px.

I again bet that my models will be different than yours. Note that in every model, ∀x Px ⇒ ∃x Px is true.

Model #1: Let P be "____ is a pizza." Let x (the universe of discourse) be all objects in the world. (In this case ∀x Px is false.)

Model #2: Let P be "____ can recite Psalm 23 by heart." Let x be all the people in the world. (In this case ∀x Px is false.)

Model #3: Let P be "Fred Gauss loves ___." Let x be all the students at KITTENS University. (In this case ∀x Px is true and ∃x Px is true.)

Model #4: Let P be "____ is a date before December 8, 1941." Let x be all possible dates after January 1, 1900. (In this case ∀x Px is false.)

Model #5: Let P be "x + x = 2x" and x be all integers. (In this case ∀x Px is true and ∃x is true.)

Model #6: Let P be "Joe ate a ___ sometime last summer." Let x be that red jelly bean that Darlene gave him at the beginning of last summer. (In this case, the universe of discourse has only one item.) (In this case ∀x Px is true and ∃x is true.)

Puzzle #97: How did I get $\mathscr{T}$ = ∃x(x = c & ∃y(diag(x) = y & $\mathscr{S}$(y)))?

I took line 3. Let $\mathscr{T}$ be the sentence ∃x(x = c & $\mathscr{C}$) and replaced the $\mathscr{C}$ by its definition in line 1: Let $\mathscr{C}$ be the string ∃y(diag(x) = y & $\mathscr{S}$(y)).

Puzzle #98: Prove ⊨ ((P & ¬Q) ⇒ Q) ⇒ (P ⇒ Q).

P	Q	¬Q	P&¬Q	(P&¬Q) ⇒ Q	P ⇒ Q	((P&¬Q) ⇒ Q) ⇒ (P ⇒ Q)
T	**T**	**F**	**F**	**T**	**T**	**T**
T	**F**	**T**	**T**	**F**	**F**	**T**
F	**T**	**F**	**F**	**T**	**T**	**T**
F	**F**	**T**	**F**	**T**	**T**	**T**

Puzzle #99: Where is the mistake in Fred's original presentation of the Diagonal lemma and the two Incompleteness Theorems?

I gave a hint in the page leading up to the **GLOOM**. I wrote:

> For the proof of the First Theorem, the Diagonal lemma implied that there is a sentence $\mathcal{G}$ with Gödel number g such that $\mathcal{G} \Leftrightarrow \neg$provable(g).
>
> Then there were two cases: Either $\mathcal{G}$ and $\neg$provable(g) are both **T**, or they are both **F**. If they were both **T**, then the proof was over.
>
> If they were both **F**, then that leads to the mess of $\mathcal{G}$ being both **T** and **F**.

Let's look at that last sentence: "If they were both **F**, then that leads to the mess of $\mathcal{G}$ being both **T** and **F**."

We didn't like the thought of $\mathcal{G}$ being both **T** and **F**. That was intolerable.

Why?

We were making the assumption at this point that the arithmetic model was consistent. If it weren't consistent, then $\mathcal{G}$ being both **T** and **F** would be just fine. It would be something that Duck would like.

Hidden in the proof of the First Incompleteness Theorem was the assumption that the model we were looking at was consistent.

Here is the correct statements of the Incompleteness Theorems:

Gödel's First Incompleteness Theorem: In any *consistent* theory that includes the Peano postulates, there are true statements that can't be proved.

Gödel's Second Incompleteness Theorem: In any *consistent* theory that includes the Peano postulates, we will never be able to prove consistency inside the model.

Index

Index

Index

To learn more
about all the books
in the
Life of Fred series

visit
FredGauss.com